MACHINERY MANAGEMENT

How to select machinery to fit the real needs of farm managers

Farm Business Management

PUBLISHER
DEERE & COMPANY
JOHN DEERE PUBLISHING
One John Deere Place
Moline, IL 61265
http://www.johndeere.com/publications
1–800–522–7448

Farm Business Management is a series of manuals created by Deere & Company. Each book in the series is conceived, researched, outlined, edited, and published by Deere & Company, John Deere Publishing. Authors are selected to provide a basic technical manuscript that could be edited and rewritten by staff editors.

HOW TO USE THE MANUAL: This manual can be used by anyone — experienced mechanics, shop trainees, vocational students, and lay readers.

Persons not familiar with the topics discussed in this book should begin with Chapter 1 and then study the chapters in sequence. The experienced person can find what is needed on the "Contents" page.

This current EDITION was produced by the technical writers, illustrators, and editors of Almon, Inc. — a full-service technical publications company headquartered in Waukesha, Wisconsin (www.almoninc.com).

FOR MORE INFORMATION: This book is one of many books published on agricultural and related subjects. For more information or to request a FREE CATALOG, call 1–800–522–7448 or send your request to address above or:

Visit Us on the Internet
http://www.johndeere.com/
publications

AUTHORS: John C. Siemens, Ph.D., Professor of Agricultural Engineering, University of Illinois, has authored numerous publications and educational materials on tillage systems for corn and soybean production, oil compaction, and farm machinery selection and management. Dr. Siemans has spent 31 years at the University of Illinois in extension, research, and teaching.

Wendell Bowers, professor emeritus of agricultural engineering, Oklahoma State University, has authored books, numerous publications and educational materials on machinery management. Professor Bowers, 30 years of experience in machinery management includes 7 years as a consultant in six foreign countries.

CONSULTING EDITOR: Robert G. Holmes, Ph.D., Professor of Agriculture Engineering, The Ohio State University, has had 30 years of experience in research, design, teaching, and consulting related to agricultural machinery. Professor Holmes has authored many publications related to specialty crop mechanization, soil compaction, and machinery management.

SPECIAL ACKNOWLEDGEMENTS: The authors and editor wish to thank the following people for their assistance: Richard G. Nelson, Ph.D., Extension Energy Specialist, Kansas State University; Lon R. Shell, Ph.D., Professor of Agricultural Science, Southwest Texas State University, and the following Deere & Company personnel: Frank M. Zoz, Product Engineering Center; Lyle E. Stephens, Richard R. Johnson, and Jack C. Wiley, Technical Center; Dawn C. Chard, Technical Information; and Tom Leonard, JDIS.

CONTRIBUTORS: The following groups were helpful in providing technical information and literature: Ford New Holland, New Holland, PA; Great Plains Manufacturing Incorporated, Assaria, KS; Illinois Farm Bureau, Bloomington, IL; Successful Farming, Des Moines, IA; Tractor Testing Program, Institute of Agriculture and Natural Resources, University of Nebraska; Agricultural Engineering Department, University of Arizona; Agricultural Engineering Department, University of Illinois; and the American Society of Agricultural Engineers.

PUBLISHER: Farm Business Management (FBM) texts and visuals are published by John Deere Publishing, ALMON-TIAC Building, Suite 140, 1300 19th Street, East Moline, IL 61244. This text and its supporting materials are intended only as an educational media and should not be considered a substitute for operator's manuals for specific operating procedures and safety precautions.

FOR MORE INFORMATION

This text is part of a complete series of texts and visuals on agricultural management entitled Farm Business Management (FBM). An instructor's kit is also available for each subject. For information, request a free catalog of educational materials. Send your request to John Deere Publishing, ALMON-TIAC Building, Suite 140, 1300 19th Street, East Moline, IL 61244 or call 1-800-522-7448.

 We have a
long-range interest in
Your Farming Success

Copyright © 1975, 1981, 1987, 1992, 1999, 2008. Litho in U.S.A. DEERE & COMPANY, Moline, IL/Sixth Edition/All rights reserved.   ISBN 0-86691-350-5

This material is the property of Deere & Company, John Deere Publishing, all use and/or reproduction not specifically authorized by Deere & Company, John Deere Publishing is prohibited.

I MATCHING MACHINES AND POWER UNITS

1 INTRODUCTION

Introduction 1-3
Typical Problems 1-4
How Can Operating Costs Be Kept to a Minimum? 1-6
Why Study Machinery Management? 1-6
Test Yourself 1-7
      Questions 1-7

2 MEASURING MACHINE CAPACITY

Introduction 2-1
Chapter Objectives 2-2
Capacity Measuring Methods 2-2
      Field Capacity 2-2
      Material Capacity 2-3
      Throughput Capacity 2-3
Machine Capacity 2-3
Selecting the Best Operating Speed 2-4
Theoretical Field Capacity 2-8
Effective Field Capacity 2-9
      Selecting Machine Size 2-11
      Determining Labor Needs 2-11
Metric Equivalents 2-11
Metric Equivalent Weight per Hour 2-11
Summary 2-12
Test Yourself 2-13
      Questions 2-13

3 IMPROVING FIELD EFFICIENCY

Introduction 3-1
Chapter Objectives 3-2
      Unused Capacity 3-2
      Filling Procedures Efficiencies 3-2
      Unloading Procedures 3-3
      Turning Time and Field Conditions 3-3
      Unclogging Machines 3-4
      Making Adjustments 3-5
      Reducing Breakdowns 3-5
      Servicing Machines 3-5
      Rest Stops 3-5
      Changing Operators 3-6
      Checking Machine Performance 3-6
      Unmatched Machine Capacity 3-6
      Estimated Field Efficiencies 3-7
Summary 3-7
Test Yourself 3-8
      Questions 3-8

4 MATCHING MACHINE SIZE AND CAPACITY

Introduction 4-1
Chapter Objectives 4-1
    Estimating Effective Field Capacity 4-2
    Fitting Capacity to Time Available 4-5
Summary 4-5
Test Yourself 4-7
    Questions 4-7

5 ESTIMATING POWER REQUIREMENTS

Introduction 5-1
Chapter Objectives 5-2
    Engine Types 5-2
    Power Ratings 5-2
    Soil Resistance to Machines 5-6
    Tractor Sizes 5-7
Summary 5-13
Test Yourself 5-14
    Questions 5-14

II ESTIMATING MACHINERY COSTS

6 ESTIMATING FIXED COSTS

Introduction 6-3
Chapter Objectives 6-3
Depreciation 6-4
    Straight-Line Depreciation 6-5
    Sum-of-the-Digits Depreciation 6-6
    Declining-Balance Depreciation 6-7
Other Fixed Costs 6-9
    Taxes 6-10
    Shelter 6-10
    Insurance 6-10
    Interest 6-10
Estimating Fixed Costs 6-11
    Low-Annual-Use Fixed Costs 6-11
How to Reduce Fixed Costs 6-12
Summary 6-14
Test Yourself 6-14
    Questions 6-14

7 ESTIMATING FUEL AND LUBRICANT COSTS

- Introduction 7-1
- Chapter Objectives 7-2
- Estimating Fuel Needs for Crop Production 7-2
  - Horsepower-Hours of Energy 7-2
  - Types of Fuel 7-4
  - Comparing Fuel Consumption of Cropping Systems 7-4
- Estimating Average Fuel Consumption 7-5
  - Estimating Average Fuel Consumption for Tractors 7-5
  - Estimating Fuel Consumption for Self-Propelled Machines 7-6
- Estimating Average Fuel and Lubricant Costs 7-7
  - Estimating Lubricant Costs 7-7
  - Estimating Both Fuel and Lubricant Costs 7-7
- Fuel-Saving Tips 7-8
- Metric Equivalents 7-9
  - Calculating Fuel and Lubricant Costs in Metric Units 7-9
- Summary 7-10
- Test Yourself 7-11
  - Questions 7-11

8 ESTIMATING REPAIR COSTS

- Introduction 8-1
- Chapter Objectives 8-2
- Types of Repairs 8-3
  - Routine Wear 8-3
  - Accidental Breakage or Damage 8-3
  - Repairs Due to Operator Neglect 8-4
  - Routine Overhauls 8-4
- Calculating Lost-Time Costs 8-4
- Establishing Life of Equipment 8-5
- Estimating Repair Costs 8-6
- Summary 8-9
- Test Yourself 8-10
  - Questions 8-10

9 TOTAL COSTS FOR MACHINES AND OPERATIONS

- Introduction 9-1
- Chapter Objective 9-2
- Cost for Individual Machines 9-2
  - Using Machine Cost Tables 9-2
  - All Wheel-Type Tractors 9-3
- Estimating Machine Cost 9-4
  - Estimating Tractor-Machine Costs 9-6
  - Chisel Plows, Mulch Tillers, Disks, Cultivators, Harrows, etc. 9-6
  - Estimating Costs for Combines 9-8
  - System Costs 9-9
- Summary 9-11
  - Why It Is Important to Know Total Cost 9-11
- Test Yourself 9-11
  - Questions 9-11

III MANAGING MACHINERY

10 DECIDING WHEN TO TRADE

- Introduction 10-3
- Chapter Objectives 10-3
- Establishing Trading Guidelines 10-3
  - Average Cost per Unit of Use 10-4
  - Machine Obsolescence 10-6
  - Machine Reliability 10-6
  - Worn-Out Machinery 10-6
  - Calculating Machine Life 10-7
  - When to Repair 10-7
- Summary 10-9
- Test Yourself 10-9
  - Questions 10-9

11 CONSIDERING FUTURE CAPACITY NEEDS

- Introduction 11-1
- Chapter Objectives 11-2
- Selecting Tractor Size 11-2
  - Calculating Relative Costs 11-3
  - Costs for 100-Horsepower (75-kW) Tractor 11-3
  - All Wheel-Type Tractors 11-4
  - Costs for 125-Horsepower (93-kW) Tractor 11-5
  - Allowing for Expansion 11-6
- Selecting Machine Size 11-7
  - Timeliness 11-7
  - Putting a Value on Timeliness 11-9
- Metric Equivalents 11-10
  - Selecting Tractor Size 11-10
- Summary 11-10
- Test Yourself 11-11
  - Questions 11-11

12 CALCULATING CUSTOM WORK COSTS

- Introduction 12-1
- Chapter Objectives 12-2
  - Determining and Comparing Costs 12-2
  - Determining Annual Fixed Cost 12-3
  - Determining Average Operating Costs 12-4
  - Determining Total Costs 12-4
  - Doing Custom Work to Help Justify Ownership Cost 12-5
- Establishing a Rate for Custom Work 12-5
  - Calculating Amount of Custom Work Needed to Justify Ownership . 12-7
- Adding Custom Work to Reduce Costs 12-8
- Metric Equivalents 12-9
  - Determining and Comparing Cost 12-9
- Summary 12-9
- Test Yourself 12-10
  - Questions 12-10

13 DECISION TIME — SELECTING THE BEST ALTERNATIVE

Introduction 13-1
Chapter Objective 13-1
What Are the Alternatives? 13-1
    Renting 13-1
    Comparing Alternatives 13-3
    Comparing Two Alternatives 13-5
    Comparing Four Alternatives 13-8
Summary 13-10
Test Yourself 13-10
    Questions 13-10

14 CASE STUDIES IN MACHINERY MANAGEMENT

Introduction 14-1
Chapter Objective 14-1
Case Study — Record Keeping 14-1
Case Study — Cost of New Machinery 14-4
Case Study — Financial Analysis 14-5
    Peer Pressure 14-5
Case Study — Using a Budget 14-6
Case Study — Cash Flow Analysis 14-7
Summary 14-8
Test Yourself 14-8
    Questions 14-8

15 APPENDIX

Fixed and Repair Cost Tables A-2
    All Wheel-Type Tractors A-2
    Crawler Tractors A-3
    Self-Propelled Combines Including Grain Heads A-4
    Cotton Harvesters A-5
    Planters and Drills A-6
    Moldboard Plows A-7
    Chisel Plows, Mulch Tillers, Disks, Cultivators, Harrows, etc. A-8
    Mowers A-9
    Large and Small Square Balers A-10
    Large Round Balers A-11
    Self-Propelled Forage Harvesters A-12
    Self-Propelled Windrowers A-13
    Rakes A-14
Explanation of Worksheet A-19
Nebraska OECD Tractor Test 1775 — Summary 310 .... A-21
    John Deere 8310 Diesel, 16 Speed A-21
Parameters for Estimating Repair Costs A-27
Weights and Measures A-28
Measurement Conversion Chart A-28
    Acreage Chart for Various Row Lengths and Implement Widths A-29
    Seeds or Plants per Acre, Thousands A-30
    Determining Hectares for Various Row Lengths and Implement Widths A-30
    Seeds or Plants per Hectare, Thousands A-31
Probabilities for a Working Day A-32
Suggested Readings A-33
Glossary of Terms A-34

Matching Machines and Power Units

	Chapter
Introduction	1
Measuring Machine Capacity	2
Improving Field Efficiency	3
Matching Machine Size and Capacity	4
Estimating Power Requirements	5

Preface

The ability to manage is an important skill that must be mastered by farmers and ranchers who want to compete in our complex worldwide commodity marketplace. With the basic information in this book, you can build a solid foundation of knowledge that can be used to make the most efficient machinery management decisions and help keep your business competitive.

There is no substitute for personally solving a machinery management problem. You should not rely on the opinions or "shortcut" methods of others when you make decisions that affect your business. The data and formulas in this text are based on the latest information available at the time of publication. Adapting this information to your own situation will greatly improve the accuracy of your machinery management decisions.

The book is divided into three sections. Section One covers the topic of how to become more efficient by matching machines and power units to different situations. Section Two gives information that can be used to estimate and analyze costs so that better machinery management decisions can be made. Section Three gives several examples of the application of information contained in the first two sections to illustrate the value of making decisions on a sound, economical basis.

Throughout this book, emphasis is placed on solving problems with a calculator or computer. Procedures will be shown in the Appendix for any problems throughout the book where a calculator or computer can be used. You will be pleasantly surprised at the wide variety of management decisions that can be made, once you master these example problems.

Matching Machines and Power Units

The first step in becoming a skilled machinery management decision-maker is to learn how to properly select and match power units. You must choose the most economical complement of tractors and machinery to achieve the desired results for each farm or ranch operation.

The first section consists of five chapters. This information will help you estimate:

- Machine capacity
- Field efficiency
- Machine capacity requirements
- Power requirements

The second section provides procedures to estimate machinery costs, including:

- Fixed costs
- Fuel and lubrication costs
- Repair costs
- Total costs

The third section will add to your abilities to make machinery management decisions. The section covers the following topics:

- When to trade
- Future machinery needs
- Custom work
- Renting and leasing
- Case studies

You will find the Appendix especially useful. The Appendix includes tables for estimating fixed costs and repair costs for all common farm machinery. There are also practice problems to solve using a calculator or computer. The Appendix concludes with list of suggested readings, a glossary of terms, and an index.

Introduction

Fig. 1 — Modern Agricultural Machinery Must Be Skillfully Managed for Maximum Profits

Introduction

Machinery management has increased in importance in today's farming operations because of its direct relation to the success of management in mixing land, labor, and capital to return a satisfactory profit (Fig. 1). Machinery costs include fixed (ownership) costs and variable (operating) costs.

The importance of machinery in the total farming operation is indicated by the machinery costs in relation to the total costs. Typically, machinery costs overshadow all other crop production costs except land. Machinery costs often account for 50% of total production costs (Fig. 2), and can run as high as $200 per acre ($500 per hectare) per year for intensive cropping systems on irrigated land. It is not unusual to find that the difference in profit from one farm to another is due solely to differences in the machinery selected and the way it is managed.

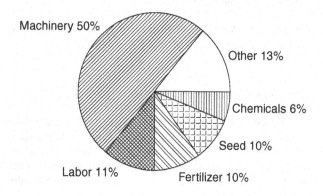

Fig. 2 — Excluding Land Costs, Machinery Costs Overshadow All Other Costs. Because They Change, Current Rates Should Be Reviewed. University Cooperative Extension Is a Good Resource

Typical Problems

The problems of machinery management are quite varied. Following are some typical examples of the more important problems. In each case the decision could mean the difference between a profit and a loss.

- How much machinery should be owned?
- What size equipment is needed?
- Should a custom operator be used?
- How often should machinery be traded?
- Should a machine be repaired or traded?

These are only five examples of the many important decisions related to owning and operating agricultural machinery.

How Much Machinery Should Be Owned?

A modern farm usually involves production of several different crops, with each one having its own tillage, planting, and harvesting requirements. You could make up quite a list of special tools by size and type for each field operation. Each machine purchased means more overhead costs — depreciation, taxes, shelter, insurance, and interest. If these fixed costs are excessive, they can rapidly eat up profits.

Lack of adequate equipment may mean not getting crops planted or harvested on time. The delay of key operations results in a timeliness loss.

Efficient selection and management of the equipment for your farm or ranch should help you achieve both short and long range financial goals.

What Size Equipment Is Needed?

Larger machines have lower labor costs. Larger tractors, for example, furnish plenty of power to complete big acreage jobs in a hurry (Fig. 3). But unless the tractor has a high annual use or you plan to own it for 10 years or more, overhead (or fixed costs) can exceed even the most expensive labor costs.

Fig. 3 — Big Tractors Provide More Capacity, but Need Lots of Use to Hold Down Costs

Smaller tractors cost less per hour than larger tractors. By having high annual use, you may be ready to trade in 6 to 8 years instead of 10 to 12 years (Fig. 4). Smaller tractors have less capacity which, in turn, can cause delays in key field operations.

Fig. 4 — Smaller Tractors Have Less Capacity, but May Have Lower Cost per Acre

By carefully analyzing the work to be done in the available time, tractors can be selected to work with the correctly sized machines to complete important field operations on time.

Size selection of machinery is based on a combination of expected performance and expected costs. Both capacity and capital costs increase with size. At the same time, performance improves, particularly with critical operations such as planting and harvesting. Delays in planting can reduce yields. Delays in harvesting can reduce both quantity and quality of production. These are timeliness losses.

Fig. 5 illustrates the interaction of machinery, labor, and timeliness costs to help determine the optimum size of machinery. Chapter 11 provides more information on machine size and timeliness costs.

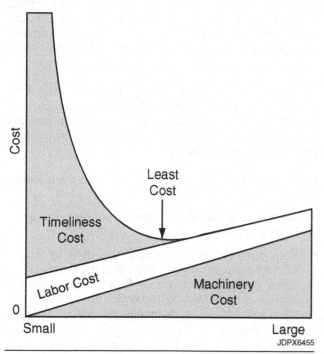

Fig. 5 — *Optimum Machine Size Considers Costs for Machinery, Labor, and Timeliness*

Should a Custom Operator Be Used?

There may be situations where it is either more practical or less expensive to use a custom operator (Fig. 6).

Fig. 6 — *For Some Situations, Hiring a Custom Operator Costs Less Than Ownership*

If you are growing a specialty crop, particularly with a small acreage, it may be more practical to use a custom operator. The custom operator can justify ownership of expensive specialty machines by the combined acreage of several different producers.

In some situations, the cost of using a custom operator may be less than your cost of ownership. In addition, the capacity of custom operators with multiple machines may be considerably higher than for your own machine.

In the final analysis, the cost benefit of using a custom operator needs to be weighed against the advantage of owning.

How Often Should Machinery Be Traded?

Frequent trading, in some cases, means always having the latest and best equipment. Having new equipment generally reduces downtime and repairs, but may result in a higher cost per acre.

Some managers must trade machinery more often because they neglect their maintenance program. Trading often can increase the average cost. With good maintenance, most agricultural machines last 8 to 10 years or more, producing the lowest unit cost.

Older machines are more likely to have a higher percentage of downtime and can easily become obsolete. Repair costs also increase with the hours of use.

To know exactly when to trade a machine, it is important to know how to take into account these four elements:

- Average cost per unit of use
- Cost to trade compared to making repairs
- Reliability of the machine
- Net effect on profit of total farming operation

Should a Machine Be Repaired or Traded?

The decision of whether you should repair the existing machine or trade for a new one may be difficult to make, unless you have an accurate estimate of how it will affect profits for the total farming operation. The age of the current machine and the history of the machine's reliability are important factors (Fig. 7).

Fig. 7 — When to Trade or Repair Is a Major Decision

If a machine has a recent history of breakdowns, then it may have reached the end of its useful life. Owning machinery that is reliable and meets quality and performance needs in the field is one of the best ways to hold down the cost of production.

How Can Operating Costs Be Kept to a Minimum?

Operating costs include labor, fuel, lubricants, and repairs. In comparison to fixed costs such as depreciation, taxes, housing, insurance, and interest; operating costs are usually lower. However, you can control operating costs. This manual explains estimating operating costs and keeping them to a minimum without reducing production.

Reducing operating costs by 10% to 15% is a realistic goal that could help you save 5 dollars or more an acre each year.

Improperly matching tractors and implements can increase operating costs as much as 15% to 20% in wasted fuel. If it is necessary to use tractors at partial loading, gear up and throttle back to save as much as 50% on fuel consumption.

Why Study Machinery Management?

In the days of farming with horses and even when many farms had their first tractor, machinery-related decisions were of less importance. The farms were small with few machines, and most of the management methods were routine.

Machinery management is a complex subject. You must be able to solve difficult machinery management problems in order to help maximize returns.

You must learn as much as possible about correct management of agricultural machinery to maximize your profits. While you may never know if your decision was the best one, at least you should have eliminated costly mistakes.

No matter the size of the farm, it is difficult to maximize returns when commodity prices are low and input costs are high. While you may not have much control over prices, you can do something about input costs resulting from machine purchases and maintenance.

Some suggestions for preparing for machinery management are:

- Learn how to use the management principles presented in this manual.
- Keep complete records of field work done by various machines and the number of working days available for critical field operations (Fig. 8). By knowing the average capacity of machines and the number of work days available, you can do a more effective job of selecting machinery.

Fig. 8 — Complete and Accurate Records Pave the Way to More Profitable Decisions

- Know how to accurately estimate costs for any machine and how to combine costs of machines to estimate total cost for an entire system. Many important decisions are based on costs.
- Know how to improve equipment reliability — always work toward the elimination of unnecessary downtime.

Introduction

- Improve field efficiencies with machines to cut costs and complete more work in the available time (Fig. 9).

Fig. 9 — Look for Ways to Improve Machinery Efficiency

- Develop short and long range plans for your farming operation, including the repair, purchase, and trade-in of equipment.

- Think of ways to improve the efficient ownership and management of agricultural machinery. It is unlikely a single decision will produce instant financial success. With the proper approach, however, you will learn how to arrive at a series of proper decisions that will increase net profit.

- Speed up management decisions by using a calculator or computer to review the problems.

Throughout this book, paragraphs and formulas containing only metric terminology are noted by the symbol. Ⓜ

Test Yourself

Questions

1. Which of the farming costs is the most likely to exceed costs for machinery?

2. Give three specific examples of machinery management decisions that affect profits.

3. Which tractor would have the highest percent of its annual costs as fixed costs: a large tractor used 400 hours a year or a small tractor used 800 hours a year?

4. How many years should most agricultural machinery last with proper care and maintenance if it has typical annual use?

5. What are three ways that average farm managers can improve their ability to make the proper machinery management decisions?

6. What is the machinery manager's goal when buying machinery?

7. What are the two alternatives available when deciding to buy machinery?

8. What are three things that can be done to hold down operating costs?

9. List four or more operations in which timeliness is critical. What would happen in each case if the operation were not completed on time?

10. If machinery costs are $125 per acre each year and, through good management, the farmer saves 15% of these costs, what annual savings ($/acre) would result?

Measuring Machine Capacity

2

Introduction

The capacity of a machine is its rate of performance. Depending on the kind of machine, the performance or capacity will be measured in terms of acres per hour (hectares per hour), tons per hour (metric tons per hour), bushels per day, or hundredweight per hour (quintals per hour).

It is important for efficient farm managers to understand how to estimate capacities of machines they plan to buy for future use (Fig. 1).

Fig. 1 — Estimating Capacity Is Important in Order to Match Machine Size to Available Working Time

Chapter Objectives

- Calculate theoretical machinery capacity using U.S. customary or metric units.
- Calculate effective machinery capacity.
- Match power unit to machinery width and speed.
- Select machinery capacity to match available time for planting, harvesting, or tillage.
- Determine labor needs based on the effective field capacity.

It is important to know machine capacities for selection of power units and equipment that can complete important field operations on time (Fig. 2). But, it is also important to avoid the added expenses of larger-than-necessary machines.

Fig. 2 — Adequate Planting Capacity Is Important to Secure Maximum Yields

Capacity Measuring Methods

The capacity of a machine is its rate of performance, usually reported in terms of quantity per time. Acres per hour (hectares per hour) is the most common measure of machine performance. Harvesting and processing machine performances may be measured as bushels or tons per hour. Bushels per hour is a rather inaccurate measure due to variations in moisture content and grain densities. In many cases the hundredweight (cwt) measure, which is 100 pounds, is used. In metric units, the common units include kilograms, metric tons (1000 kg), and quintals (100 kg).

Capacity of harvesting machines, in some cases, requires measurements other than area per unit of time.

Three possible measurers of machine capacity include:

- Field capacity in acres or hectares per hour
- Material capacity in hundredweight (cwt) or kilograms per hour
- Throughput capacity in pounds, tons, kilograms, or metric tons per hour

Field capacity, usually expressed in acres (hectares) per hour, is the most commonly used measure of machine capacity.

Material capacity, once commonly referred to in bushels per hour, is now usually measured by hundredweight (cwt) or kilograms per hour. In many cases, tons per hour is used. Material capacity is simply a measure of material, such as silage or grain, harvested by a machine. But, in the case of harvesting machines such as combines, this is not a completely accurate measure of capacity, because it measures only the amount of grain or forage harvested. To measure total amount of material passing through the machine, throughput capacity is used.

Throughput capacity is a special term used to compare machines, such as combines and potato harvesters, that separate undesirable material from the desirable material. In these cases, the weight of the material handled is the accurate capacity measure. Throughput, then, refers to the time rate of handling a total weight of material, usually in terms of pounds (kilograms) per hour. In the case of a combine, the pounds-per-hour (kilograms-per-hour) throughput would include grain, chaff, straw, and any other material that enters the header. Because moisture affects the result, a moisture report should accompany the throughput capacity rating.

Let's consider a sample problem using these three different capacity measurements. It is determined that a combine with a 20-foot (6.1-meter) wide header is combining wheat at a speed of 300 feet (92 meters) per minute. In a 1-minute time period, 1000 pounds (454 kilograms) of material enters the header. Of this amount, 500 pounds (227 kilograms) enters the grain tank and the remaining 500 pounds (227 kilograms) of material is discharged through the combine.

Use the unit-factor method to determine the different capacities.

Field Capacity

Speed times width equals field capacity:

$$\frac{300 \text{ feet}}{\text{minute}} \times \frac{60 \text{ minutes}}{1 \text{ hour}} \times 20 \text{ feet} \times \frac{1 \text{ acre}}{43{,}560 \text{ ft}^2}$$

Answer = 8.26 acres per hour

Units of measure in the numerator cancel the identical units of measure in the denominator. Apply the unit-factor method to the equations on the following pages.

Material Capacity

Pounds (kilograms) harvested per hour:

$$\frac{500 \; \cancel{\text{cwt}}}{\cancel{\text{minute}}} \times \frac{60 \; \cancel{\text{minutes}}}{1 \; \text{hour}} \times \frac{1 \; \text{cwt}}{100 \; \cancel{\text{pounds}}}$$

Answer = 300 cwt (or hundredweight) per hour

This answer could also be expressed in tons per hour:

$$\frac{300 \; \cancel{\text{cwt}}}{\text{hour}} \times \frac{100 \; \cancel{\text{pounds}}}{1 \; \cancel{\text{cwt}}} \times \frac{1 \; \text{ton}}{2{,}000 \; \cancel{\text{pounds}}}$$

Answer = 15 tons per hour

Throughput Capacity

Total amount of material per hour:

$$\frac{1{,}000 \; \text{lb}}{\cancel{\text{minute}}} \times \frac{60 \; \cancel{\text{minutes}}}{1 \; \text{hour}} = 60{,}000 \; \text{pounds per hour}$$
$$= 30 \; \text{tons per hour}$$

These examples are briefly covered here to familiarize you with the technical capacities. For most practical farm or ranch applications, acres per hour and tons per hour are the preferred ways of measuring capacity. Material capacity requires accurate measuring techniques and is easily affected by moisture, yields, and other factors such as platform cutting height.

All of these capacities are theoretical capacities, not effective capacities. Effective capacity brings in the factor of efficiency. After reviewing practical methods of measuring machine capacity, the factor of efficiency will be introduced.

NOTE: See the Appendix for conversion tables for weights and measures.

Machine Capacity

The units of measure of machine capacity used throughout this manual will be:

- Acres per hour (hectares per hour)
- Tons per hour (U.S. customary and metric)

Learning to use these terms is important for proper management.

Field Capacity

Field capacity, when measured in acres per hour (hectares per hour), is determined by three factors:

- Speed
- Width
- Efficiency

Speed is the average rate of travel expressed in miles per hour (mph) or kilometers per hour (km/h).

Width is the distance in feet (meters) across the processing portion of the machine.

Efficiency is the ratio of the effective field capacity of a machine to its theoretical field capacity. It is the ratio of how much time is spent working and total time in field.

Material Capacity

Capacity of a machine like a forage harvester or a silage blower is often expressed in tons per hour (Fig. 3). The only major difference is the use of weight as a unit of work accomplished instead of area.

Fig. 3 — Capacity Is Measured in Tons per Hour for Some Machines

Let's take a closer look at each of the factors that determine capacity and learn how to make critical management decisions.

Selecting the Best Operating Speed

Most field machines work best at a given speed. For example, modern primary tillage tools (Fig. 4) work best at speeds of 4 to 6 miles per hour (6.4 to 9.7 kilometers per hour). Going too slow will not allow the tillage tool to provide enough breaking of the soil. Going too fast might give too much shattering or throw the soil too far. Power may be wasted due to higher draft at increased speeds.

Fig. 4 — There Is a Preferred Speed Range for Each Field Operation

The speed may need to be quite slow for some operations, such as cultivating small row crops, to avoid damage to the plants. On the other hand, modern rotary hoes need to run fast to kick out weeds (Fig. 5). A rotary hoe operating at speeds from 6 to 10 miles per hour (9.7 to 16.1 kilometers per hour) is not unusual.

Fig. 5 — In Some Cases, a Faster Speed Is Necessary to Make an Implement Work Properly

There are also operations, such as chopping heavy-yielding silage or combining high-yielding grain, where the density of the crop being harvested calls for extra power. In such cases, the extra power requirement usually limits speed.

Recent engineering trends have resulted in farm machinery being designed to operate at higher speeds. This is particularly true for tillage tools and, in some cases, hay equipment. This subject is discussed in more detail in Chapter 5. With an increase in speed, greater emphasis is needed on safety, not only for the operator, but also for the equipment.

Speed is measured in miles per hour (mph) or kilometers per hour (km/h). Some operations, like spraying, may require a precise speed if the sprayer is not equipped with a rate controller (Fig. 6).

Fig. 6 — A Precise Speed Is Needed to Ensure Proper Application Rate in Many Spraying Operations

It is a good idea to make an accurate speed check because speed indicators may be affected by tire size, field conditions, and other factors (Fig. 7). If your tractor or machine is equipped with radar or similar device (Fig. 8) to give actual ground speed, then it may not be necessary to make a calibration check.

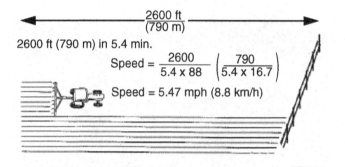

Fig. 7 — One Method of Checking Speed When You Know the Field Length

Fig. 8 — Some Tractors Are Equipped With Radar to Measure Speed Accurately

The procedure for making an accurate speed check is not difficult. Just determine how far the tractor or machine travels in a given period of time.

Suppose you check the speed for 1 minute and find the tractor traveled exactly 440 feet (134 meters). In 1 hour, a distance 60 times as far could be traveled because there are 60 minutes to an hour:

$$60 \frac{\text{min.}}{\text{hr}} \times 440 \frac{\text{ft}}{\text{min.}} = 26{,}400 \frac{\text{ft}}{\text{hr}}$$

(M) $60 \frac{\text{min.}}{\text{hr}} \times 134 \frac{\text{meters}}{\text{min.}} = 8{,}040 \frac{\text{meters}}{\text{hr}}$

Because there are 5,280 feet in 1 mile, the speed would be 26,400 divided by 5,280, or 5 miles per hour.

There are 1,000 meters in 1 kilometer. Divide 8,040 by 1,000 to obtain 8.04 kilometers per hour.

There are two convenient methods or formulas for checking speed. The more accurate method is shown in Fig. 7. Determine the exact length of the field (or distance traveled) and the time needed to travel the distance in minutes. The formula is as follows:

$$\text{Speed, mph} = \frac{\text{distance, feet}}{\text{time, minutes}} \times \frac{60 \text{ minutes}}{1 \text{ hour}} \times \frac{1 \text{ mile}}{5{,}280 \text{ feet}}$$

(M) $\text{Speed, km/h} =$

$$\frac{\text{distance, meters}}{\text{time, minutes}} \times \frac{60 \text{ minutes}}{1 \text{ hour}} \times \frac{1 \text{ kilometer}}{1{,}000 \text{ meters}}$$

Then the equation for making a speed check is as follows:

$$\text{Speed, mph} = \frac{\text{distance, feet}}{\text{time, minutes} \times 88}$$

(M) $\text{Speed, km/h} = \dfrac{\text{distance, meters}}{\text{time, minutes} \times 16.7}$

If, as shown in Fig. 7, you drive 2,600 feet (792.5 m) in 5.4 minutes, the speed would be calculated as follows:

$$\text{Speed, mph} = \frac{2{,}600}{5.4 \times 88} = 5.47 \text{ mph}$$

(M) $\text{Speed, km/h} = \dfrac{792.5}{5.4 \times 16.7} = 8.8 \text{ km/h}$

An easier and simpler method of measuring speed is to mark off a distance of 88 feet (or 16.7 meters) in the field. Then, with a running start, check the number of seconds needed to drive between the markers using the rear axle on the tractor as a reference point (Fig. 9). The formula for calculating speed when using this method is:

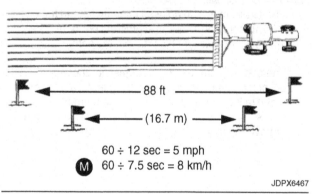

60 ÷ 12 sec = 5 mph
(M) 60 ÷ 7.5 sec = 8 km/h

Fig. 9 — To Make a Simple but Accurate Speed Check, Divide 60 by the Time in Seconds to Travel 88 Feet (or 16.7 Meters)

$$\text{Speed, mph} = \frac{60}{\text{seconds to travel 88 ft}}$$

(M) $\text{Speed, km/h} = \dfrac{60}{\text{seconds to travel 16.7 meters}}$

Measuring Machine Capacity

Let's try an example. It takes 11 seconds to drive 88 feet (6.8 seconds to drive 16.7 meters) with a self-propelled swather. What is the speed?

Speed, mph = $\frac{60}{11}$ = 5.45 mph

(M) Speed, mph = $\frac{60}{6.8}$ = 8.8 km/h

TIME-SPEED RELATIONS*	
Time to Travel 88 Feet, Seconds	Speed, mph
3	20.0
4	15.0
5	12.0
6	10.0
7	8.6
8	7.5
9	6.7
10	6.0
11	5.5
12	5.0
13	4.6
14	4.3
15	4.0
16	3.8
18	3.3
20	3.0
24	2.5
28	2.1
32	1.9
36	1.7
40	1.5

Table 1 — Time-Speed Relations
 * This table may also be used to relate time to travel 16.7 meters and speed in kilometers per hour. The number and relationships remain the same.

This method is also useful in adjusting the speed to a desired point. Suppose you want to operate a planter at a speed of exactly 5 miles per hour. The same method may be used, except now divide 60 by the desired speed to get the time to travel 88 feet (or 16.7 meters) (Fig. 10):

Required time (seconds) = $\frac{60}{\text{speed}}$

For 5 mph, time = $\frac{60}{5}$ or 12 seconds

(M) For 8 km/h, time = $\frac{60}{8}$ or 7.5 seconds

Fig. 10 — To Plant at 5 Miles per Hour (8 km/h), Adjust the Speed Until You Drive 88 Feet in 12 Seconds (or 16.7 Meters in 7.5 Seconds)

Adjusting the speed to make the run of 88 feet in exactly 12 seconds (or 16.7 meters in 7.5 seconds) ensures a planting speed of 5 miles per hour (8 km/h).

In order to allow operation of different types of equipment at the proper speed, manufacturers provide tractors with a wide range of field speeds. This is accomplished in two different ways. As shown in Fig. 11, a multi-range transmission provides as many as 16 forward and 6 reverse speeds.

Fig. 11 — Modern Tractors Have a Wide Selection of Speeds

Another type of transmission is hydrostatic drive. It allows operation at the exact speed that allows a machine to perform at its best. The windrower shown in Fig. 12 uses hydrostatic drive. The operator can vary the speed from one moment to the next to compensate for any variations of crop conditions in the field. Machine speed is determined the same way as with other transmission types.

Fig. 12 — Hydrostatic Drive Lets You Work at the Best Speed for Various Field Conditions

Increasing Average Machine Width

Using full machine width is one important way to more efficiently use labor and equipment. The greater the average width of cut, the greater the capacity (Fig. 13). Every machine should be used as close to its full width as possible.

Fig. 13 — Unused Cutting Width Reduces Field Efficiency and Capacity

If a 14-foot (4.27-meter) self-propelled windrower is operated at an average width of cut of 13 feet 2 inches (4 meters): 10 inches (0.25 meters) or nearly 6% of the available width is lost. Although a skilled operator probably cannot cut a full 14 feet (4.27 meters), experience with field conditions indicates that an average cutting width of 13 feet 6 inches (4.1 meters) is easily possible. Skilled operators might average closer to 13 feet 8 inches (4.16 meters), but they have to watch closely where they are driving.

Theoretical Field Capacity

Theoretical field capacity is the maximum possible capacity obtainable at a given speed, assuming the machine is using its full width (Fig. 14).

Theoretical field capacity is explained in the following example. Suppose a tractor pulls a 14-foot (4.27-meter) wide disk at 5.0 mph (8.0 km/h). What is the theoretical capacity?

The distance traveled in 1 hour would be:

$$5.0 \frac{mi}{hr} \times 5{,}280 \frac{ft}{mi} = 26{,}400 \frac{feet}{hr}$$

Ⓜ $8.0 \frac{km}{hr} \times 1{,}000 \frac{meters}{km} = 8{,}000 \frac{meters}{hr}$

Assuming the disk always cuts the full width of 14 feet (4.27 meters), it would cover:

$$14\ ft \times 26{,}400 \frac{ft}{mi} = 369{,}000\ \text{square feet per hour}$$

Ⓜ $4.27\ m \times 8{,}000 \frac{m}{h} = 34{,}160\ \text{square meters per hour}$

Because there are 43,560 square feet in an acre (10,000 square meters in a hectare), the theoretical field capacity of the disk is:

$$\frac{369{,}600}{43{,}560} = 8.48\ \text{acres per hour}$$

Ⓜ $\frac{34{,}160}{10{,}000} = 3.42\ \text{hectares per hour}$

Theoretical field capacity, TFC, can also be determined with the following formula:

$$\text{TFC, acres/hr} = \frac{\text{Speed, mph} \times \text{Width, ft}}{8.25}$$

Fig. 14 — Theoretical Capacity Is an Instantaneous Capacity for a Given Speed

Let's derive the constant 8.25 using the unit-factor system. We want to calculate theoretical field capacity in units of acres per hour when we know speed and width of an implement. We already know that theoretical field capacity in acres per hour =

$$\frac{\text{Square ft covered per hr}}{43{,}560\ \text{square ft/acre}}$$

We also know that square feet covered per hour =

Speed, mph x Width, ft x 5,280 ft/mi

Therefore, theoretical field capacity in acres per hour =

Speed, mph x Width, ft x 5,280 ft x acres/43,560 ft^2

Next divide 43,560 by 5,280 and we get the equation for theoretical field capacity in acres per hour:

$$\text{TFC, acres/hr} = \frac{\text{Speed, mph} \times \text{Width, ft}}{8.25}$$

Remember, when using the equation, speed must be in miles per hour and width in feet.

Now let's use the equation to determine the theoretical field capacity of the disk. The disk is 14 feet wide and is pulled at a speed of 5.0 miles per hour.

$$\text{TFC, acres/hr} = \frac{5 \times 14}{8.25} = 8.5 \text{ acres/hr}$$

See Metric Equivalents later in this chapter for the derivation of the equation and determination of theoretical field capacity of the disk in metric units.

Theoretical field capacity cannot be sustained for long periods of time because any field operation will be interrupted by turns, filling hoppers, and breakdowns (Fig. 15). Yet, it is valuable information. It gives the maximum capacity, which can be used as a basis for evaluating performance of machines and their operators. The effective field capacity is always less than the theoretical capacity.

Fig. 15 — Time Lost in Filling Hoppers, Making Turns, and Other Operations Reduces Effective Field Capacity

Effective Field Capacity

The best way to determine the effective field capacity (EFC) of a machine is to make an accurate check of the area actually covered or weight handled over a long period of time.

EFC in Area per Hour

If a 14-foot (4.27-meter) disk actually covers 70 acres (28.3 hectares) while operating for 10 hours with no breakdowns, its effective field capacity would be:

70 acres divided by 10 hours, or 7.0 acres per hour

 28.3 hectares divided by 10 hours, or 2.83 hectares per hour

But one day's experience may not give a true picture of the effective field capacity for the season (Fig. 16). If the same disk is used for a two-calendar-week period, we might have different figures such as:

Total calendar days	= 14
Total working days	= 8
Total hours in field	= 72
Total acres (hectares) covered	= 480 (195)
Effective field capacity	= $\frac{\text{Total acres}}{\text{Total hours}}$ = $\frac{480}{72}$
	= 6.67 acres per hour

(M) $\frac{\text{Total hectares}}{\text{Total hours}} = \frac{195}{72} = 2.71$ hectares per hour

Now let's go back and compare the field efficiency for the one-day period and the two-week period.

In the one-day period, the capacity was 7.0 acres per hour (2.83 hectares per hour):

$$\text{Field efficiency} = \frac{\text{Effective field capacity}}{\text{Theoretical capacity}} \times 100$$

$$= \frac{7.0}{8.48} \times 100 = 83\%$$

(M) $\frac{2.83}{3.42} \times 100 = 83\%$

But for the two-week period, the field efficiency would be:

$$\text{Field efficiency} = \frac{6.67}{8.48} \times 100 = 79\%$$

(M) $\text{Field efficiency} = \frac{2.71}{3.42} \times 100 = 79\%$

Thus, the field efficiency for the two-week period was lower than for the one-day check.

Fig. 16 — Use Realistic Field Capacity Based on Average for the Season

EFC in Weight per Hour

The effective field capacity for hay and forage equipment is essentially calculated in the same way as effective field capacity in units of area per hour. In the calculations, weight is substituted as the unit of measure in the place of area.

To obtain EFC in metric tons per hour, see the section under the heading Metric Equivalent Weight per Hour.

If a self-propelled forage harvester is chopping windrowed alfalfa that yields 1 ton per acre, its effective field capacity could be expressed in tons per hour. If 100 acres of the alfalfa is chopped in 10 hours with no breakdowns or other delays, its effective field capacity would be:

$$\frac{1 \text{ ton/acre} \times 100 \text{ acres}}{10 \text{ hours}} = 10 \text{ tons per hour}$$

As in the previous example, one day's experience may not give a true picture of the effective field capacity for the season. If the same forage harvester is used for a four-calendar-week period, different figures might evolve such as:

Total calendar days = 28
Total working days = 18
Total hours in field = 180
Total tons (metric tons) harvested = 1600 (1445)

$$\text{Effective field capacity} = \frac{\text{Total tons}}{\text{Total hours}} = \frac{1600}{180}$$

$$= 8.89 \text{ tons per hour}$$

Now let's compare the field efficiency for a one-day period and the four-week period.

In the one-day run, the capacity was 10 tons per hour. Assume theoretical capacity was calculated to be 12.5 tons per hour.

$$\text{Field efficiency} = \frac{\text{Effective field capacity}}{\text{Theoretical capacity}} \times 100\%$$

$$= \frac{10}{12.5} \times 100\% = 80\%$$

But for the four-week period, the field efficiency would be:

$$\text{Field efficiency} = \frac{8.89}{12.5} \times 100\% = 71\%$$

Selecting Machine Size

Machine size can be determined by estimating the time available and determining the speed of the operation. Using field capacity to determine the size of a machine is a good management practice.

$$\text{Width} = \frac{\text{acres per hour} \times 8.25}{\text{miles per hour}}$$

Determining Labor Needs

Another reason to determine machine efficiency is to estimate labor needs. If hired labor is needed for a field operation, it is useful to calculate the amount and cost of the labor needed. As a management tool, machine capacity can be used to determine the expenses such as labor cost.

Consider this problem: A producer can hire a person at $11.00 per hour. The effective field capacity for cultivating is 10 acres per hour and you have 200 acres to till. The cost of labor needed would be $220.00.

$$\text{Labor cost} = \frac{200 \text{ acres}}{10 \text{ acres per hour}} \times \$11.00 \text{ per hour}$$

Metric Equivalents

Sample problems shown in the beginning of the chapter. A combine with a header 6.1 meters wide travels at a speed of 92 meters per minute. In 1 minute, 454 kg of material enters the header. Of this material, 227 kg enters the grain tank and the remaining 227 kg is discharged through the combine. Determine the theoretical field capacity, material capacity, and throughput capacity.

Theoretical field capacity equals speed times width:

$$\frac{92 \text{ meters}}{\text{minute}} \times \frac{60 \text{ minutes}}{1 \text{ hour}} \times 6.1 \text{ meters} \times \frac{1 \text{ hectare}}{10,000 \text{ m}^2}$$

$$= 3.37 \text{ hectares per hour}$$

Material capacity equals kilograms harvested per hour:

$$\frac{227 \text{ kilograms}}{\text{minute}} \times \frac{60 \text{ minutes}}{1 \text{ hour}} = 13,620 \text{ kg per hour}$$

Material capacity in metric tons per hour equals:

$$\frac{13,620 \text{ kilograms}}{\text{hour}} \times \frac{1 \text{ ton}}{1,000 \text{ kilograms}} = 13.62 \text{ tons per hour}$$

Total amount of material per hour:

$$\frac{454 \text{ kilograms}}{\text{minute}} \times \frac{60 \text{ minutes}}{1 \text{ hour}} = 27,240 \text{ kilograms per hour}$$

$$= 27.24 \text{ tons per hour}$$

Theoretical field capacity, TFC, can be determined with the following formula:

$$\text{TFC, ha/hr} = \frac{\text{Speed, km/h} \times \text{Width, m}}{10}$$

Let's derive the constant 10 using the unit-factor system. We want to calculate theoretical field capacity in units of hectares per hour when we know speed and width of an implement. We already know that theoretical field capacity in hectares per hour =

$$\frac{\text{Square meters covered per hour}}{10,000 \text{ square meters/hectare}}$$

We also know that square meters covered per hour =

Speed, km/h × Width, m × 1,000 m/km

Therefore, theoretical field capacity in hectares per hour =

Speed, km/h × Width, m × 1,000 m/km × ha/10,000 m^2

Next divide 10,000 by 1,000 and we get the equation for theoretical field capacity in hectares per hour:

$$\text{TFC, ha/hr} = \frac{\text{Speed, km/h} \times \text{Width, m}}{10}$$

Remember, when using the equation, speed must be in kilometers per hour and width in meters.

Now let's use the equation to determine the theoretical capacity of the disk. The disk is 4.27 meters wide and is pulled at a speed of 8.0 kilometers per hour.

$$\text{TFC, ha/hr} = \frac{8.0 \times 4.27}{10} = 3.42 \text{ ha/h}$$

Metric Equivalent Weight per Hour

A metric ton is 1,000 kilograms or 2,205 pounds. One hectare = 10,000 square meters = 2.47 acres. One kilogram = 2.205 pounds. One pound = 0.4536 kilograms. One meter = 100 centimeters.

A forage harvester that can average 15 tons (U.S. customary) per hour could harvest 150 tons in a 10-hour day.

The harvest in 10 hours would be 150 tons x 2,000 lb/ton or 300,000 pounds.

Since the metric ton is 2,205 pounds, 300,000 lb ÷ 2,205 lb/metric ton = 136 metric tons per day.

If the yield is 5 tons per acre (U.S. customary), the yield in metric units would be 5 tons/acre x 2,000 lb/ton x 2.47 acres/ha = 24,700 lb/ha.

24,700 lb/ha ÷ 2,205 lb/metric ton = 11.2 tons (metric)/ha

Summary

Measuring machine capacity is basic to machinery management. Machine capacity is the machine's rate of performance, usually reported in terms of quantity per time.

The three possible measurers of machine capacity are:
- Effective field capacity
- Material capacity
- Throughput capacity

Throughout this manual we will use field capacity, measured in acres per hour (hectares per hour), and material capacity, measured in pounds (kilograms) or tons per hour, as the two most common methods of measuring machine capacity. Other methods of measuring machine capacity are generally too technical for common farm application.

Speed, width, and efficiency are the three factors used to determine effective field capacity.

Theoretical field capacity uses two of the factors — speed and width. It is the maximum possible capacity obtainable at a given speed, assuming the machine is using its full width.

Effective field capacity brings in the factor of efficiency. This capacity determination represents the real-life or actual capacity obtainable over a period of time.

A comparison of the one-day and two-week field efficiencies in this chapter shows a mistake commonly made by farm operators or managers. The mistake is the natural tendency to remember only the best working days and to use these capacity figures for planning and scheduling purposes. When scheduling operations or sizing machines for the future, use realistic field capacities as determined over the full season of use (Fig. 16). Use accurate capacities for the same machine in different operations, such as disking stalks and smoothing soil prior to planting.

Test Yourself

Questions

1. Name the three factors that determine capacity of a machine.

2. List three or four operations where it is important for a machine to run at the proper speed.

3. If you are disking in a field 1,056 feet long and travel from one end to the other in 2.4 minutes, the speed in miles per hour is:

   a. 0.5 mph

   b. 5.0 mph

   c. 6.5 mph

4. You want to adjust a tractor with a mounted sprayer to run 4.6 miles per hour. How many seconds should it take to travel 88 feet?

5. (T/F) The theoretical field capacity of a combine with a 20-foot header traveling 4.0 miles per hour is 9.7 acres per hour.

6. If the theoretical field capacity of a 14-foot self-propelled windrower is seven acres per hour and it cuts 63.7 acres in 13 working hours, what is the field efficiency?

   a. 70%

   b. 41%

   c. 91%

7. What is the effective field capacity of a 6-row, 30-inch header equipped corn combine traveling 3.33 miles per hour, assuming a field efficiency of 60%?

8. (Fill in blank.) You have a broadcast sprayer that needs to be calibrated. In a speed check, it travels 88 feet in 12 seconds. The speed is _____ miles per hour.

9. What is the field efficiency of a 16-foot chisel plow traveling an average of 4.5 miles per hour if it averages 65 acres in a 10-hour day?

   a. 82.5%

   b. 74.5%

   c. 65%

10. You need to estimate the effective capacity of a corn planter that you are thinking of buying. Describe the four steps you would follow.

11. If your tractor travels 400 meters in 4 minutes, what is the speed in kilometers per hour?

    a. 6.0 km/h

    b. 9.7 km/h

    c. 5.1 km/h

12. A combine with a 6-row corn head adjusted for 76-centimeter row spacing harvests at 6 kilometers per hour. What is the theoretical field capacity in hectares per hour?

13. If field efficiency is 62%, what is the effective field capacity in hectares per hour?

14. (Fill in blank.) The same broadcast sprayer from question 8 once again needs to be calibrated. In this instance, it traveled 16.7 meters in 7.5 seconds. The speed is _____ kilometers per hour.

15. A combine with a 7-meter head travels 8 kilometers an hour. What is the theoretical field capacity in hectares per hour?

16. How many hectares are equal to 100 acres?

17. How many acres are equal to 1,000 hectares?

Improving Field Efficiency

Fig. 1 — A Pre-Season Checkup Improves Field Efficiency by Helping to Eliminate Breakdowns

Introduction

The ability to improve field efficiency is the next important step in developing machinery management skills. There are several important reasons why a machine may have a certain field efficiency. Some lost-time factors are built into the operation. Other lost-time factors can be eliminated by good planning and management (Fig. 1). Typical factors causing lost time include:

- Unused capacity
- Filling procedures
- Unloading procedures
- Turning and field conditions
- Unclogging machines
- Making adjustments
- Reducing breakdowns
- Servicing machines
- Rest stops
- Changing operators
- Checking machine performance
- Unmatched machine capacity

Any of these factors may contribute greatly to lost time. Let's discuss these factors and their effects on field efficiency in more detail.

Chapter Objectives

- Identify factors that may reduce theoretical field capacity.
- Identify factors that may improve field capacity efficiency.
- Create a list of daily inspections to check performance criteria for specific machinery.
- Explain the reasons for variable field capacity efficiency percentages.

Unused Capacity

The way an individual operates a machine determines how much of its potential capacity is actually used. This is true in cases of using the fullest possible width of a machine or keeping the machine adjusted for maximum capacity (Fig. 2).

Fig. 2 — Drive to Utilize Full Operating Width of Implements

A 28-foot (8.5-meter) field cultivator that overlaps 3 feet (0.9 meters) each trip over a field loses 11% of its effective capacity. A 12-foot (3.6-meter) pull-type windrower that cuts only 11 feet (3.35 meters) loses 8% of its width. A 4-bottom, 14-inch (36-cm) plow that cuts only 10 inches (25.4 cm) with the front bottom is losing 7%. A forage harvester with improperly positioned or dull knives requires more power and results in more breakdowns because of increased strain on machine components.

These are just a few examples of how improper field operation or adjustment can affect machine capacity. Proper field operation and machine adjustments result in greater capacity for any machine. Average values of efficiency losses for unused width will range from 5% to 10%, but this type of loss can be reduced to 4% or less by better planning, careful driving, and correct machine adjustment.

Filling Procedures Efficiencies

Proper filling procedures are particularly important to field efficiency because they are part of seeding or planting operations for crops. Time losses of up to 20% are not unusual for refilling hoppers on a planter or drill. If travel time is also needed to refill tanks or hoppers, the time lost may be nearer to 30%. Time between refills can be greatly reduced by using larger hoppers (Fig. 3).

Fig. 3 — Large Supply Hoppers Improve Field Efficiency

One way to improve field efficiencies with filling operations is to use larger hoppers, portable augers, and nurse tanks with transfer pumps (Fig. 4). It is important to complete the planting as quickly as possible from a timeliness standpoint. Therefore, constantly look for ways to increase field efficiency. It may be better to refill hoppers in the middle of long fields than to refill boxes that may be one-third to one-half full. With proper organization and mechanization, these time losses can be reduced by 50%.

Fig. 4 — Augering of Materials Increases Productivity

Unloading Procedures

If excessive maneuvering is required to spot trucks or wagons to unload the combine, it's not unusual to use 20% to 30% of the total field time in unloading, particularly if the unloading auger is slow. Skilled operators can cut this time in half by minimizing turning time and spotting the truck or wagon so it is conveniently located on the turning path. This lost time is virtually eliminated by unloading on-the-go (Fig. 5).

Fig. 5 — Unload On-the-Go for Maximum Field Efficiency

Turning Time and Field Conditions

Turning time is the one loss factor built into every field operation, but can be kept to a minimum by planning fields with longer rows (Fig. 6). Turning time can run as high as 25% of the total field time, but more typically the range will be from 12% to 15%.

A normal turn at the end of a field is one that can be made with one continuous motion and requires little, if any, extra motion to return to the centerline. If the space at the end of the field is too narrow for a complete turn and requires backing for completion, as much as 50% more turning time will be required.

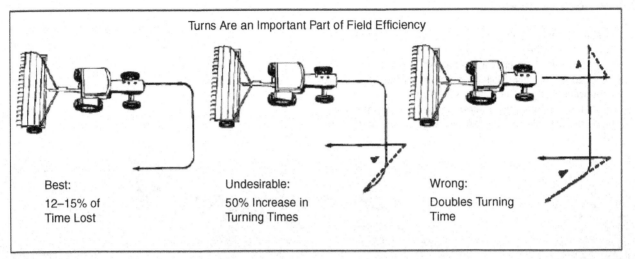

Fig. 6 — Turning Time Needs to Be Kept to a Minimum

If there is no space at the end of a field for a turn and the entire operation must be completed by backing, turning time may be more than doubled. Other time losses are caused by rough field conditions, obstructions, and ditches.

Unclogging Machines

One time factor that can be nearly eliminated by good management is unclogging machines. It is also a main reason for an unusually high percentage of lost field time and accidents. Under severe conditions, the effective field capacity may be only one-third or one-half of what it would be without clogging problems (Fig. 7).

Fig. 7 — Poor Harvest Conditions Increase Clogging, Reduce Effective Field Capacity, and May Cause Accidents

Machines clog for three reasons. The most likely reason for clogging is overloading the machine. Even if the machine is in top shape, it can still plug up if operated over its designed capacity.

Secondly, there could be something wrong with the machine, such as dull knives, faulty slip clutches, loose belts, or improper adjustment. Keeping the machine in top condition reduces excessive clogging (Fig. 8).

Fig. 8 — Keep the Machine in Proper Condition to Avoid Excessive Clogging

Finally, trying to operate the machine when conditions are not ideal can also cause clogging. The crop may be too wet or too dry. The ground may be too wet (for tillage) or there may be too much crop residue.

Making Adjustments

With most machines, it is necessary to make adjustments before going to the field (Fig. 9). Some machines, like combines or balers, have to be adjusted occasionally in the field as crop conditions vary throughout the day. For maximum performance from forage harvesters, for example, rotating knives should be kept sharp and the stationary knife or shear bar kept properly adjusted. Always consult the operator's manual for making adjustments to get the maximum performance from machines.

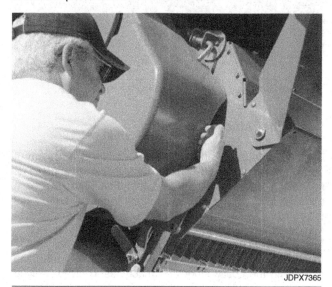

Fig. 9 — Making Adjustments Before Going to the Field Improves Field Efficiency

The worst mistake made with complex machines like combines or planters is to wait until getting to the field to make repairs and adjustments. Have the machine checked over, repaired, adjusted, and ready for operation before going to the field.

Reducing Breakdowns

It is impossible to predict when some part of a machine will fail, but many machine breakdowns in the field can be avoided by making thorough inspections before and during operation. Watch for signs that will help avoid breakdowns. To help eliminate or reduce breakdowns, follow these rules:

- Inspect and repair machines well ahead of the use season.
- Practice preventive maintenance.
- Avoid rocks, holes, and obstructions.
- Drive cautiously in rough fields.
- Don't overload the equipment.
- Check out strange sounds, vibrations, or smells.
- Make minor repairs when needed.
- Use periodic checkups to locate potential trouble.
- Keep all power-transmitting members adjusted, aligned, and lubricated.
- Safety alert! A well-running machine is a safer machine.

Servicing Machines

Most maintenance should take place before or after the day's work, but many machines require frequent lubrication during the day. Such maintenance cannot be neglected (Fig. 10). It is far better to use extra time for complete and thorough machine servicing than to have field breakdowns costing hours or even days.

Fig. 10 — Proper Servicing Helps Prevent Field Breakdowns

Rest Stops

Short but frequent breaks or rest stops are desirable and necessary from the standpoint of both safety and improved performance (Fig. 11). When the operator has to continually concentrate on a job, a short break every hour or so will increase alertness, reduce accidents, and improve the quality of work.

Fig. 11 — Rest Breaks Make You More Alert and Improve Quality of Work

On the other hand, avoid prolonged or unnecessary rest stops. They cut heavily into a day's production. Allowing an extra 20 minutes in the morning and another 20 minutes in the afternoon other than regular rest breaks can reduce effective field capacity by 7% to 8%.

Think through management decisions during rest breaks or the night before, rather than while operating a machine. In the interest of safety while operating, always devote full attention to the machine being operated.

Changing Operators

Changing operators usually applies to larger operations and requires little extra time (Fig. 12). The best way to save time and improve field efficiency is to service machines and change operators right in the field. Running in to the service yard for refueling, greasing, and other service operations means unnecessary time loss.

Fig. 12 — Changing Operators and Servicing Machines in the Field Increases Efficiency

Checking Machine Performance

Only a few minutes a day is required to stop occasionally to inspect a machine and check performance (Fig. 13). This practice, while it takes time, goes a long way toward improving field efficiency. Spot checking machine performance, looking for loose bolts or worn parts, often prevents bigger problems.

Fig. 13 — Take Time to Check Machine Performance

Unmatched Machine Capacity

One major cause of an inefficient operation is a system of unmatched machines. It is important to follow the principle of increasing capacity with each operation. That is, each machine involved in succeeding steps in the system needs to provide greater capacity than the preceding machine. If the machine providing the third step in a four-step system has a lower capacity than one of the machines involved in the first two steps, the other machines will be idle part of the time. With the costly labor and machines involved, waiting can be expensive.

A good example of such a system is a corn harvesting operation. In order to set up an example of a system that uses the maximum efficiency of the machines involved, we will use these three steps:

Step 1 — Harvest with a corn combine. Capacity is 400 bushels (10,170 kg) per hour or 4,000 bushels (101,700 kg) in a 10-hour day.

Step 2 — Haul to processing center. Potential hauling capacity is 450 bushels (11,440 kg) per hour or 4,500 bushels (114,400 kg) in a 10-hour day.

Step 3 — Processing, which includes drying and moving to storage. Drying rate is 300 bushels (7,627 kg) per hour or 4,800 bushels (122,035 kg) in 16 hours. This excess capacity allows for future expansion of production and is considered a good management procedure.

In this simplified example, the most important part of the operation is getting the corn out of the field.

Improving Field Efficiency

Each process that follows has an increasing capacity capability. This is to make sure that the steps following harvesting keep pace with the initial or primary harvesting operation. Be sure to keep the capacity of each machine in the system in mind when planning operations. Always allow adequate capacity in each machine to keep up with the capacity of the previous operation, or "bottlenecks" will result.

Estimated Field Efficiencies

Table 1 lists estimated field efficiencies for several tillage, planting, and harvesting implements. Note that each implement has a range of up to 35%. Your own efficiency will depend on such factors as climate, soil conditions, crop conditions, length of field, and your ability to maintain and use the machine.

Summary

The upper limits of field efficiencies apply to long fields with only a small time loss for turning and other essentials. Matching tillage equipment with optimal power units will result in improved field efficiency. For maximum field efficiencies, there can be no time lost for such things as unplugging or breakdowns. The better operators will improve the long-run field efficiencies by servicing machines periodically, checking adjustments, and keeping their machines in top mechanical condition. Table 1 lists ranges in field efficiencies

FIELD EFFICIENCY TABLE	
OPERATION	FIELD EFFICIENCY, PERCENT
Tillage	
Moldboard Plow	75–85%
Disk Harrow, Disk Plow	77–90%
Field Cultivator	75–85%
Spring-Tooth or Spike-Tooth Harrow	65–80%
Cultivation	
Row Crop	65–80%
Rotary Hoe	75–85%
Seeding	
Corn Planter	
1. Corn Only	60–75%
2. With Fertilizer and/or Pesticide Attachment	45–65%
Grain Drill	65–80%
Broadcast	65–70%
Harvesting	
Mower	75–85%
Rake	65–90%
Baler	65–80%
Loose Hay Stacking Wagon	65–80%
Forage Harvester	50–70%
Combine	60–75%
Corn Picker	55–70%
Cotton Picker	60–75%
Swather	70–85%
Miscellaneous	
Sprayer	55–65%

Table 1 — Field Efficiency

Test Yourself

Questions

1. If a 14-foot disk overlaps an average of 20 inches with each trip, what percentage of the width is lost?

    a. 10.0

    b. 12.0

    c. 18.0

2. Why does field efficiency increase with field length?

3. Does field efficiency stay the same if speed is doubled? Why or why not?

4. (T/F) An 8-row planter has exactly double the capacity of a 4-row planter if the speed of planting is identical.

5. List three items the average operator can do to improve field efficiency.

6. A custom operator when harvesting wheat averages 8 acres per hour. If the operator could make a 3% improvement in field efficiency, how many more acres could be harvested in a 600-hour season?

7. What difference does 30 minutes of lost time make in a 10-hour day for a planting operation that otherwise averages 10 acres an hour?

    a. 3 acres a day

    b. 5 acres a day

    c. 10 acres a day

8. For the planting operation in problem 7, how much longer would it take to plant 1,000 acres, if 30 minutes were wasted every day? Assume a 10-hour day.

9. A self-propelled combine with a 6.1-meter head averages cutting 5.8 meters with each pass. What percent loss in field efficiency is due to the overlapping, or unused width?

10. If the combine in problem 9 operates at a speed of 6.5 kilometers per hour and a field efficiency of 75%, what is the effective field capacity in hectares per hour?

    a. 2.25 ha/h

    b. 2.97 ha/h

    c. 2.75 ha/h

11. (Fill in blank.) If the field efficiency for the combine in problem 10 drops to 60%, the effective field capacity would be _____ ha/h. How much less would the combine harvest in a 10-hour day when operating at the lower field capacity?

Matching Machine Size and Capacity

4

Fig. 1 — Proper Machine Size Means Having Adequate Capacity to Complete Critical Field Operations During the Working Time Available

Introduction

Selecting machine size is an important part of machinery management. First, machines must be the optimal size for the amount of work to be done. Second, a machine must match the capacity of related equipment (Fig. 1).

In order to fit machine size to the amount of work that has to be completed in a specific length of time, it will be necessary to understand effective field capacity (Fig. 2).

In this chapter, we will also become more familiar with theoretical field capacity, effective field capacity, and field efficiency.

Chapter Objectives

- Interpret the calculations used to determine theoretical field capacity and effective field capacity.
- Explain the way variables affect theoretical field capacity for specific farm machinery.
- Create a computer spreadsheet to determine calculations of theoretical and effective field capacity.

Fig. 2 — By Knowing Available Working Time and the Amount of Work to Be Done, You Can Determine Required Capacity

Estimating Effective Field Capacity

By combining the factors of theoretical field capacity (TFC) of a machine with its field efficiency, it is possible to calculate the effective field capacity (EFC). In many places throughout this manual the following abbreviations will be used:

- EFC (Effective Field Capacity)
- TFC (Theoretical Field Capacity)

NOTE: See section at the end of this chapter for metric equivalents for the following formulas.

Suppose we have a 16-foot chisel plow and want to find both the theoretical field capacity (TFC) and the effective field capacity (EFC). The working width is 16 feet in a field 2,600 feet long at a speed of 5.5 miles per hour (Fig. 3).

Width = 16 Feet
Speed = 5.5 Miles per Hour

85% Field Efficiency
$$EFC = 10.67 \times \frac{85}{100} = 9.07 \text{ acres per hour}$$

75% Field Efficiency
$$EFC = 10.67 \times \frac{75}{100} = 8.00 \text{ acres per hour}$$

Fig. 3 — Reducing Field Efficiency From 85% to 75% Cuts Effective Capacity From 9.07 Acres per Hour to 8.00 Acres per Hour

$$TFC = \frac{\text{Width, ft} \times \text{Speed, mph}}{8.25} = \frac{16 \times 5.5}{8.25}$$

TFC = 10.67 acres per hour

Effective field capacity (EFC) can be determined by combining the formula for theoretical field capacity (TFC) and field efficiency.

$$EFC = \frac{\text{Width, ft} \times \text{Speed, mph}}{8.25} \times \frac{\text{Field Efficiency}}{100}$$

This formula is the same as:

$$EFC = TFC \times \frac{\text{Field Efficiency}}{100}$$

Because the field is rather long, we will use a field efficiency of 85%. With longer fields a higher percentage of total time in field is spent working and less time turning. So using the formula, we see that:

$$EFC = 10.67 \times \frac{85}{100} = 9.07 \text{ acres per hour}$$

In other words, the actual capacity is 85% of the theoretical capacity (Fig. 3).

Suppose the chisel plow is used in a short, irregular field. How would the reduced field efficiency change the effective field capacity? If the field efficiency drops from 85% to 75%, the new effective field capacity will be:

$$EFC = 10.67 \times \frac{75}{100} = 8.00 \text{ acres per hour}$$

Lowering the field efficiency from 85% to 75% would decrease effective capacity by 1.07 acres per hour, 10.7 acres in 10 hours, or 107 acres in 100 hours. Continually look for ways to improve field efficiency, because the resulting increased capacity will lower production costs (Fig. 4).

Fig. 4 — Keep Thinking of Ways to Improve Efficiency Even if It Means Stopping to Think It Through

Let's try another example of calculating effective field capacity. A forage harvester is chopping haylage that has been cut and windrowed by a 14-foot windrower. The forage harvester is traveling at a speed of 3.6 miles per hour in a field 2,000 feet in length, and the hay is yielding 3.6 tons per acre. First, calculate theoretical field capacity:

$$\text{TFC} = \frac{14 \times 3.6}{8.25} = 6.1 \text{ acres per hour}$$

Because we want our answers in tons per hour, multiply the acres-per-hour theoretical field capacity by the tons-per-acre yield to get theoretical material capacity:

$$\text{TMC (tons per hour)} = \frac{6.1 \text{ acres}}{\text{hour}} \times \frac{3.6 \text{ tons}}{\text{acre}}$$

$$= 22.0 \text{ tons per hour}$$

After determining that field efficiency is 70% because of the process of changing wagons after each load, calculate the effective material capacity:

$$\text{EMC} = \frac{22.0 \text{ tons}}{\text{hour}} \times \frac{70}{100} = 15.40 \text{ tons per hour}$$

Once again, by simply using different field efficiencies it is easy to see why it is important to always have the best field efficiency possible.

Finally, let's use an example of a corn combine. Assume a 6-row, 30-inch-row corn head equipped combine is working at a speed of 3.2 miles per hour. With long rows, well-organized unloading patterns, and no breakdowns, field studies indicate that a field efficiency of 70% can be achieved (Fig. 5).

$$\text{EFC} = \frac{15.0 \text{ ft} \times 3.2}{8.25} \times \frac{70}{100} = 4.07 \text{ acres/hour}$$

Fig. 5 — A 6-Row, 30-Inch Corn Combine Will Average 4.07 Acres per Hour at 3.2 Miles per Hour With a 70% Field Efficiency

$$\text{Width} = 6 \text{ rows} \times \frac{30 \text{ in.}}{\text{row}} \times \frac{1 \text{ ft}}{12 \text{ in.}} = 15 \text{ ft}$$

$$\text{EFC} = \frac{15.0 \text{ ft} \times 3.2 \text{ mph}}{8.25} \times \frac{170}{100}$$

$$= 4.07 \text{ acres per hour}$$

With a little thought and practice, you can also solve for speed, width, or field efficiency, using the following formulas.

The equations for speed, width, and field efficiency are:

$$\text{Speed, mph} = \frac{\text{EFC, acres/hr} \times 8.25}{\text{Width, ft}} \times \frac{100}{\text{Field Efficiency}}$$

$$\text{Width, ft} = \frac{\text{EFC, ac/hr}}{\text{Speed, mph}} \times 8.25 \times \frac{100}{\text{Field Efficiency}}$$

$$\text{Field Efficiency, \%} = \frac{\text{EFC, ac/hr} \times 8.25 \times 100}{\text{Speed, mph} \times \text{Width, ft}}$$

Let's try the formula for width on an example:

How wide a disk harrow is needed for an effective field capacity of 6 acres per hour (2.43 ha/h), assuming a field efficiency of 80% and a speed of 5 miles per hour (8.04 km/h) (Fig. 6)?

$$\text{Width} = \frac{6 \times 8.25}{5.0} \times \frac{100}{80} = 12.4 \text{ feet}$$

Fig. 6 — Determining Disk Width Needed to Disk 6 Acres per Hour at 5 Miles per Hour With 80% Field Efficiency

Solve by formula:

$$\text{Width} = \frac{6 \text{ acres/hr}}{5 \text{ mph}} \times 8.25 \times \frac{100}{80} = 12.4 \text{ feet}$$

See formula in Metric Equivalents later in this chapter for metric units.

Matching Machine Size to Fit Time Available

One of the best ways to determine how large a machine needs to be is to determine the necessary capacity to complete the operation within a specified calendar period (Fig. 7). The secret to success when using this method is to carefully select the calendar period and correctly estimate the number of hours you can work during that period.

Fitting Capacity to Time Available

Let's take corn planting as an example of fitting capacity to the time available. How many calendar days do you consider to be the maximum for your situation? Five? Ten? Fifteen? Twenty? Be careful not to allow too many days. Corn needs a long growing season for maximum yields.

Fig. 7 — Fitting Equipment Size to Available Working Time Is an Important Management Task

Another important consideration is the average number of working days during the calendar period.

In this sample situation, assume that a Midwestern corn grower with 600 acres of corn to plant wants to finish planting within a 20-calendar-day period. Considering weather and other delays typical of the Midwest, only about six of these days would be suitable for field work. Using an average of 10 hours per working day, how large a planter is needed? Assume 65% field efficiency, 5 miles per hour, and 30-inch row spacing.

Solution:

Step 1. How many total hours are available?

Six days at 10 hours per day equals 60 hours.

Step 2. How much capacity in acres per hour would be needed?

$$\frac{600 \text{ acres}}{60 \text{ hours}} = 10 \text{ acres per hour}$$

Step 3. By formula,

$$\text{Width} = \frac{10 \times 8.25}{5 \times 0.65} = 25.4 \text{ feet or 305 inches}$$

$$\frac{305 \text{ width inches}}{30 \text{ inches per row}} = 10.2 \text{ rows}$$

To find the number of 30-inch rows needed, divide the total planting width needed by the width of the row:

Answer: A 12-row, 30-inch planter would be needed.

Summary

Selecting the proper size machine to fit available time limits is a simple but effective method of getting important operations done on time. Keeping accurate records on days and hours available will give you some valuable information after a few years (Fig. 8).

Fig. 8 — Keep Records to Help Estimate Available Working Periods for Each Operation

See "Probabilities for a Working Day" in the Appendix for a list of probabilities for working days for various states. Check with your State Cooperative Extension Service for data applies to your area (Fig. 9).

Fig. 9 — Record Available Working Time for Important Operations for Later Use

Once the information on working days is gathered and organized, a chart similar to the one following (Table 1) can be helpful in planning operations.

	Number of Working Days for Probability Level	
Two-Week Period	50%	90%
March 21 – April 3	4.1	0.0
April 4 – April 17	5.9	1.8
April 18 – May 2	6.6	2.7
May 3 – May 16	7.6	4.3
May 17 – May 30	8.5	4.8

Table 1 — Operations Table

Combining accurate records with the ability to accurately estimate effective field capacities will let you make valuable decisions on needed machine sizes. As a result, you will soon realize a sizable reduction in timeliness losses. Your time will be well spent in keeping records to establish the average and minimum number of working days for critical field operations. In fact, your own records are the best accurate source of information on this subject.

Metric Equivalents

The formulas in metric for obtaining theoretical field capacity and effective field capacity that appear earlier in this chapter are as follows:

Ⓜ $\quad \text{TFC} = \dfrac{\text{Width, meters} \times \text{Speed, km/h}}{10}$

Example: A chisel plow has a width of 4.88 meters and operates at a speed of 8.85 kilometers per hour.

$\text{TFC} = \dfrac{4.88 \times 8.85}{10} = 4.32 \text{ ha/h}$

Ⓜ $\quad \text{EFC} = \text{TFC} \times \dfrac{\text{Field Efficiency}}{100}$

Example: Using a field efficiency of 85% with the above TFC:

$\text{EFC} = 4.32 \text{ ha/hr} \times \dfrac{85}{100} = 3.68 \text{ ha/hr}$

Formulas in metric to determine speed, width, and field efficiency are the following:

$\text{Speed, km/hr} = \dfrac{\text{EFC, ha/h} \times 10}{\text{Width, m}} \times \dfrac{100}{\text{Field Efficiency}}$

$\text{Width, m} = \dfrac{\text{EFC, ha/h} \times 10}{\text{Speed, km/h}} \times \dfrac{100}{\text{Field Efficiency}}$

$\text{Field Efficiency, \%} = \dfrac{\text{EFC, ha/h} \times 10}{\text{Speed, km/h} \times \text{Width, m}}$

Example: Determine width of disk harrow for an effective field capacity (EFC) of 2.43 ha/h, speed of 8.04 km/h, and field efficiency of 80%.

$\text{Width, m} = \dfrac{2.43 \times 10}{8.04} \times \dfrac{100}{80} = 3.78 \text{ m}$

An example that correlates with the forage harvester example earlier in this chapter:

Windrower — 4.27 meters cutting width

Forage harvester speed — 5.79 km/hr

Yield — 8.07 t/ha

Field Efficiency — 70%

Ⓜ $\quad \text{TFC} = \dfrac{4.27 \text{ m} \times 5.79 \text{ km/h}}{10} = 2.47 \text{ ha/h}$

2.47 ha/h × 8.07 t/h = 19.95 t/h

Ⓜ $\quad \text{EFC} = 19.9 \text{ t/h} \times \dfrac{70}{100} = 13.97 \text{ t/h}$

Example problem involving planter capacity. A comparable problem appears earlier in this chapter.

Planter — 6-row, 76-centimeter row spacing

Speed — 7.24 kilometers per hour

Field Efficiency — 60%

(M) Width, m = $\frac{6 \text{ rows} \times 76 \text{ cm}}{100 \text{ cm/m}}$ = 4.56 meters

(M) TFC = $\frac{4.56 \text{ m} \times 7.24 \text{ km/}}{10}$ = 3.3 ha/h

(M) EFC = 3.3 × $\frac{60}{100}$ = 1.98 ha/h

Another example is used to determine the width requirement for a disk.

EFC — 2.43 ha/h

Speed — 8.04 km/h

Field Efficiency — 80%

(M) Width = $\frac{2.43 \text{ ha/h}}{8.05 \text{ km/h}}$ × $\frac{100}{80}$ × = 3.77 meters

Test Yourself

Questions

1. List two important reasons for selecting proper machine size.

2. Under what conditions can you actually operate a machine at its theoretical field capacity?

3. What is the theoretical capacity of a 6-row, 30-inch cultivator traveling 4.12 miles per hour?

4. (Fill in blank.) The effective field capacity of a 14-foot offset disk with a field efficiency of 80% and operating at 5 miles per hour is _____ acres per hour.

5. You have 800 acres of wheat to combine in eight calendar days with an average of 6 working hours per day. What capacity in acres per hour will the combine have to average?

6. (T/F) Assuming an average speed of 4.01 miles per hour and a field efficiency of 70% the minimum header width needed for the example in problem 5 is 24 feet.

7. A cotton field is yielding 1.5 bales per acre and is planted in 40-inch rows. What is the theoretical capacity in bales per hour of a 2-row harvester traveling 3 miles per hour? If its field efficiency is 55%, what is the effective field capacity in bales per hour?

8. What is the effective field capacity in tons per hour of a 2-row (40-inch rows) forage harvester if it travels 4 miles per hour and has a field efficiency of 60%? Yield is 15 tons per acre.

    a. 29.1 tons per hour

    b. 35.3 tons per hour

    c. 48.5 tons per hour

9. Theoretical capacity of a baler is five 70-pound bales per minute. If the field efficiency is 70%, what is the effective field capacity in tons per hour?

    a. 10.5 tons per hour

    b. 220.0 tons per hour

    c. 7.35 tons per hour

10. A corn combine averages 4,000 bushels in 10 working hours. If the theoretical field capacity is 6 acres per hour and the yield is 100 bushels per acre, what is the field efficiency of the combine?

    a. 82.5%

    b. 66.7%

    c. 60.0%

11. What is the effective field capacity in metric tons per hour of a 2-row (100-cm rows) forage harvester if it travels 6.5 km/h and has a field efficiency of 60%? Yield is 33.6 metric tons per hectare.

    a. 47.4 t/h

    b. 26.2 t/h

    c. 35.3 t/h

12. Select a planter with 76.2-cm row spacing to plant 600 hectares in 10 working days. Assume an average of 9.4 working hours per working day and a speed of 10 km/h. Assume field efficiency equals 75%. How many rows would the planter have?

    a. 8

    b. 6

    c. 12

    d. 16

Estimating Power Requirements

5

Fig. 1 — Power Units Must Be Carefully Matched to Equipment, Regardless of Size

Introduction

A major task facing modern farmers and ranchers is to match power units to the size and type of machines so all field operations can be carried out on time with a minimum of cost. It is important to match the power needs of the equipment to that supplied by the power unit (Fig. 1).

If the tractor is oversized for implements, the costs will be excessive for the work done. If the implements selected are too large for the tractor, the quality or quantity of the work may be lessened or the tractor will be overloaded, usually causing expensive breakdowns.

For more detailed information on tractor nomenclature, operation, maintenance, and safety see John Deere's Fundamentals of Machine Operation manual TRACTORS.

Some factors to consider when selecting a power unit:

- Engine types
- Power ratings
- Soil resistance to machines
- Tractor sizes
- Converting power
- Matching implements
- Determining wheel slip
- Sizing for critical work

Chapter Objectives

- Calculate drawbar horsepower.
- Calculate PTO horsepower.
- Explain how soil textures and soil conditions affect resistance of machinery.
- Determine factors used to select the power unit to operate machinery.

Engine Types

The three general types of engines currently in use include diesel, LP-gas, and gasoline. All three types are classified as internal combustion engines (Fig. 2). The diesel engine is a compression-ignition engine with only air being compressed in the cylinder. Diesel fuel is then injected into the cylinder and the fuel-air mixture is ignited by the heat of compression.

Fig. 2 — Engines Convert Fuel to Mechanical Power

Gasoline and LP-gas engines are spark-ignition engines. In both cases, the fuel-air mixture is drawn into the cylinder and ignited by a spark.

The combustion process in the cylinder converts the energy contained in fuel to a rotating power source (Fig. 3). This rotating power can be further converted into three forms:

- Drawbar pull
- PTO output
- Hydraulic system output

In some applications, like chopping forage, the power is used for pulling a load, operating a PTO shaft (Fig. 4), and hydraulically raising and lowering the forage chopper head.

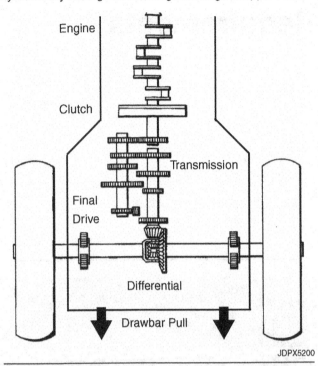

Fig. 3 — Tractors Convert Rotating Power of the Crankshaft to Drawbar Pull

Fig. 4 — Tractors Also Convert the Rotary Motion of the Crankshaft to the Output of the PTO Shaft

Power Ratings

Power is a measure of the rate at which work is being done. The U.S. customary power unit of horsepower is defined as 550 foot-pounds of work per second. The metric power unit is measured in kilowatts (kW). One kilowatt is equal to 1.34 horsepower. See Appendix, "Weights and Measures, Measurement Conversion Chart," Metric to English, for other metric equivalents.

If a load has enough weight and resistance to movement to require a force of 20 pounds to move it vertically a distance of 3 feet, the amount of work done is:

Work = force x distance (Fig. 5)
= 20 pounds x 3 feet = 60 foot-pounds

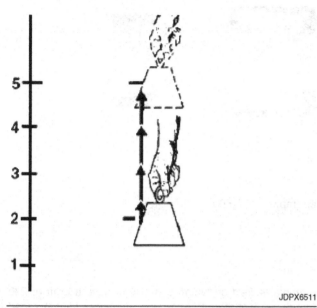

Fig. 5 — Work Equals Force Times Distance Without a Time Factor

The amount of work done is 60 foot-pounds with no reference to time.

It wouldn't matter whether the time to move the load was 1 minute or 1 hour; the amount of work stays the same. When the rate of doing the work is considered, power can be determined.

If a 1000-pound force is moved 33 feet in 1 minute, the rate of doing work is 1 horsepower, because 1 horsepower equals 33,000 foot-pounds per minute (Fig. 6). The equivalent rate of work in 1 second to equal 1 horsepower is:

$$1 \text{ horsepower} = \frac{33,000 \text{ foot-pounds}}{60 \text{ seconds}}$$
$$= 550 \text{ foot-pounds per second}$$

When working with field machinery, we usually think of speed in miles per hour and draft (or force) in pounds. For these conditions the formula for horsepower is:

$$\text{Power, hp} = \frac{\text{Force, pounds} \times \text{Speed, mph}}{375}$$

M The formula for power in kilowatts can be found in Metric Equivalents later in this chapter.

Example: If the draft of a trailing implement, like a disk harrow, is measured at 2,500 pounds and is pulled at a speed of 5 mph, what is the drawbar horsepower (Fig. 7)?

Fig. 6 — Horsepower Is a Measure of the Rate at Which Work Is Done

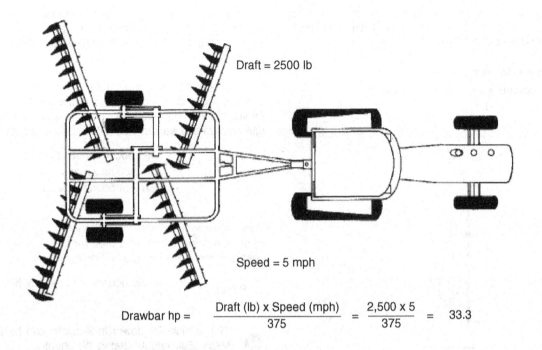

Fig. 7 — Drawbar Horsepower Can Be Calculated if Speed and Draft Are Known

$$\text{Drawbar power, hp} = \frac{\text{Force, pounds} \times \text{Speed, mph}}{375}$$

$$= \frac{2{,}500 \times 5}{375} = 33.3 \text{ hp}$$

Notice we have used drawbar horsepower. Later in the chapter we will discuss how this relates to other kinds of horsepower, such as PTO horsepower and engine horsepower.

This formula can also be used to determine how fast an implement could be pulled with a given size of tractor (Fig. 8).

$$\text{Speed, mph} = \frac{\text{Drawbar power, hp} \times 375}{\text{Draft, pounds}}$$

Example: A tractor is pulling a plow with a total draft of 5,000 pounds. How fast can the plow be pulled if the tractor has 65 drawbar horsepower available?

$$\text{Speed} = \frac{65 \times 375}{5{,}000} = 4.88 \text{ mph}$$

This formula can also be used to determine how large an implement can be pulled, but an extra step or two is involved. In the first place, the size of the implement would have to be related to the amount of soil resistance. This resistance is usually given as pounds per foot of width or pounds per square inch of soil disturbed. The latter is usually listed for moldboard plows.

Example: A field cultivator is known to have a draft of 280 pounds per foot of width when used in a given field. A speed of 6 mph is desired. The available power unit has a 130 horsepower rating at the drawbar for continuous use. What width of field cultivator could be pulled (Fig. 9)?

Step 1. Determine maximum draft for 130 horsepower at 6 mph.

$$\text{Draft, lb} = \frac{\text{Drawbar, power, hp} \times 375}{\text{Speed, mph}}$$

$$= \frac{130 \times 375}{6} = 8{,}125 \text{ lb}$$

Step 2. Determine width.

$$\text{Width, feet} = \frac{\text{Draft, lb}}{\text{Draft per foot}} = \frac{8{,}125}{280} = 29 \text{ feet}$$

From a practical standpoint, we would probably select a 28- or 29-foot field cultivator.

Estimating Power Requirements

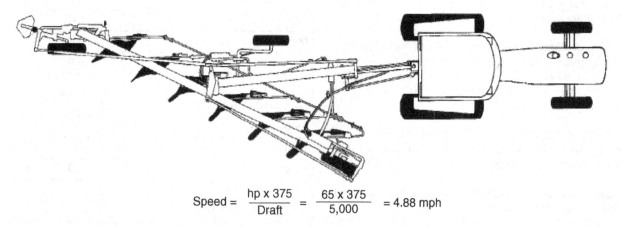

Fig. 8 — You Can Calculate Maximum Speed if the Draft and Available Horsepower Are Known

Fig. 9 — Determining Width of Implement for a 130 hp Tractor

Estimating Power Requirements

Soil Resistance to Machines

Table 1 — Soil Resistance, adapted from the Agricultural Engineering Yearbook, lists some unit draft ranges for those implements having the highest power requirements per foot of width. Table 1 also shows the horsepower required per foot of width at a typical field speed for the middle of the draft range. See Appendix, Table 16, for metric equivalents.

It is best to size power units for the maximum draft that might be encountered and at a speed for optimum machine performance. As a result, you will have a power reserve for the tough spots in a field and avoid the possibility of overloading the power train.

Table 1 can be used to determine required drawbar power. Suppose we want to determine how much drawbar horsepower is required to pull a 16-foot chisel plow (Fig. 12).

The table can be used directly to determine required drawbar power. Let's try an example.

We want to determine how much drawbar horsepower is required to pull a 16-foot chisel plow (Fig. 10).

Speed = 5.5 Miles per Hour
Width = 16 Feet
Draft = 500 Pounds per Foot

Fig. 10 — How Much Drawbar Horsepower Would Be Needed to Pull a 16-ft Chisel Plow at 5.5 mph if Soil Resistance is 500 lb/ft?

Table 1 shows a typical speed 5.5 mph for medium draft conditions of 500 pounds per foot of width. At 5.5 miles per hour, 7.3 horsepower per foot would be needed.

Multiplying total width by horsepower per foot gives total horsepower required:

16 ft x 7.3 horsepower per foot = 116.8 hp

Tillage Tool/Soil Type	Soil Resistance		
	Draft, lb/ft	Typical Speed, mph	Drawbar Power, hp/ft
COULTER — CHISEL PLOW			
Fine	575	5.0	7.7
Medium	500	5.5	7.3
Coarse	400	6.0	6.4
MOLDBOARD PLOW			
Fine	1,200	4.5	14.4
Medium	920	5.0	12.3
Coarse	600	5.0	8.0
FIELD CULTIVATOR			
Fine	390	5.0	5.2
Medium	340	5.5	5.0
Coarse	270	6.0	4.3
TANDEM DISK HARROW			
Fine	400	4.5	4.8
Medium	340	5.0	4.5
Coarse	300	5.5	4.4
OFFSET OR HEAVY DISK			
Fine	525	4.5	6.3
Medium	460	5.0	6.1
Coarse	430	5.0	5.7
ONE-WAY DISK			
Fine	400	4.5	4.8
Medium	300	5.0	4.0
Coarse	200	5.0	2.7
V-RIPPER			
Fine*	2,400	4.0	25.6
Medium*	1,900	4.25	21.5
Coarse*	1,200	4.5	14.4

Table 1 — Soil Resistance
* Draft, lb/shank; Drawbar power, hp/shank

The table could also be used to determine how large an implement could be pulled with a given size of tractor. Suppose you have a tractor with 85 usable drawbar horsepower. How large a plow will it pull in a loam or medium soil?

Table 1 shows 12.3 horsepower per foot of width. Dividing 85 by 12.3 we get 6.9 feet of width (Fig. 11).

Fig. 11 — At 12.3 Horsepower per Foot, 85 Usable Drawbar hp Could Pull 5-Bottom, 16-Inch Plow

6.9 ft x 12 in./ft = 82.8 inches

Thus, it would pull a 5-bottom, 16-inch plow. The table is intended only as a guide, because draft will vary considerably, according to soil conditions and depth of operation.

Tractor Sizes

Even when it is known how much power is needed for a given field operation, knowing the size of tractor to use still presents a problem.

First, there are several different kinds of power measurements, all applying to the same tractor. For a better understanding of the drawbar-power and tractor-size relationship, let's start at the engine and define the various kinds of power:

- Brake
- PTO
- Drawbar

Remember, tractor power is measured in horsepower (the customary measurement in the U.S.A.) or in kilowatts (the metric equivalent).

Brake power is the maximum power the engine can develop without alterations. The engine, before installation in a tractor, can be connected to a dynamometer to determine how much brake power can be developed (Fig. 12). This figure is particularly useful in sizing stationary engines for operating irrigation pumps, grinders, and other equipment. The same engines used for large tractors are often used as stationary engines.

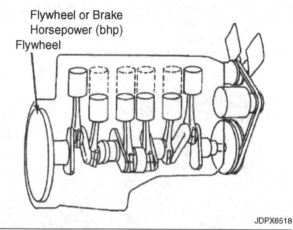

Fig. 12 — Brake Horsepower Is a Measurement of the Power Developed at the Flywheel

Power takeoff horsepower, PTO hp for short, is the power measured at the PTO shaft (Fig. 13).

POWER TAKE-OFF PERFORMANCE

Power HP (kW)	Crank shaft speed rpm	Gal/hr (l/h)	lb/hp.hr (kg/kW.h)	Hp.hr/gal (kW.h/l)	Mean Atmospheric Conditions
MAXIMUM POWER AND FUEL CONSUMPTION					
Rated Engine Speed—(PTO speed—1041 rpm)					
97.18 (72.47)	2300	6.18 (23.39)	0.444 (0.270)	15.73 (3.10)	
Standard Power Take-off Speed (PTO speed - 1000 rpm)					
100.86 (75.21)	2209	6.18 (23.40)	0.428 (0.260)	16.32 (3.21)	
Maximum Power (2 hours)					
103.99 (77.55)	1950	5.85 (22.14)	0.393 (0.239)	17.78 (3.50)	
VARYING POWER AND FUEL CONSUMPTION					
97.18 (72.47)	2300	6.18 (23.39)	0.444 (0.270)	15.73 (3.10)	Air temperature
86.41 (64.43)	2401	5.80 (21.96)	0.469 (0.285)	14.90 (2.93)	79°F (26°C)
65.33 (48.72)	2425	4.90 (18.54)	0.523 (0.318)	13.34 (2.63)	Relative humidity
43.97 (32.79)	2459	4.00 (15.13)	0.634 (0.386)	11.00 (2.17)	50%
22.25 (16.59)	2460	3.09 (11.71)	0.971 (0.590)	7.19 (1.42)	Barometer
2.12 (1.58)	2460	2.19 (8.30)	7.220 (4.392)	0.97 (0.19)	29.11" Hg (98.58 kPa)

Maximum Torque - 330 lb.-ft. (448 Nm) at 1300 rpm
Maximum Torque Rise - 48.7%

Fig. 13 — Maximum PTO Power Is the Most Commonly Used Method to Rate Tractor Power (from a Nebraska Tractor Test Report)

Drawbar power is a measure of the pulling power of the engine by way of tracks, wheels, or tires. As a percentage of PTO power, the drawbar power varies, depending on several factors including soil surface and type of hitch. As shown earlier, drawbar power is a function of drawbar pull and speed.

Of the various methods of rating tractor power, rated PTO power is the one most commonly used, as being measured in Fig. 14. Exceptions are some of the larger four-wheel drive tractors and track-type tractors that do not have a PTO shaft. They may be given an engine power (flywheel) or maximum drawbar power rating (Fig. 15).

Fig. 14 — PTO Power Being Measured by a Shop Dynamometer

DRAWBAR PERFORMANCE
UNBALLASTED - FRONT DRIVE ENGAGED
FUEL CONSUMPTION CHARACTERISTICS

Power Hp (kW)	Drawbar pull lbs (kN)	Speed mph (km/h)	Crank-shaft speed rpm	Slip %	Fuel Consumption lb/hp.hr (kg/kW.h)	Hp.hr/gal (kW.h/l)	Temp.°F (°C) cooling med	Air dry bulb
Maximum Power—7th (B3) Gear								
87.84 (65.51)	7740 (34.43)	4.26 (6.85)	2295	3.23	0.495 (0.301)	14.10 (2.78)	184 (84)	80 (27)
75% of Pull at Maximum Power—7th (B3) Gear								
69.28 (51.66)	5793 (25.77)	4.49 (7.22)	2403	2.59	0.555 (0.337)	12.59 (2.48)	183 (84)	79 (26)
50% of Pull at Maximum Power—7th (B3) Gear								
47.34 (35.30)	3867 (17.20)	4.59 (7.39)	2435	1.63	0.664 (0.404)	10.51 (2.07)	178 (81)	82 (28)

JDPX6521

Fig. 15 — Maximum Drawbar Power Is Another Method of Rating Tractor Size (from a Nebraska Tractor Test Report)

So the problem really becomes one of knowing how much power any given tractor will have available for constant use on the drawbar. This problem is further complicated by different kinds of soil conditions. The less firm the soil, the more power lost at the drawbar. Speed, tire size, slippage, and ballasting are all important factors in obtaining the best operating conditions for a specific situation.

Converting Power

A factor of 87% can be used to convert either engine or drawbar power ratings to an equivalent rated PTO power rating.

For example, a tractor with a rating of 350 engine horsepower would have 87% of 350, or 304.5, as the equivalent rated PTO horsepower.

If another tractor is rated at 200 drawbar horsepower, as tested on concrete, the drawbar horsepower would be approximately 87% of the equivalent rated PTO power. Dividing 200 by 0.87 gives a rated PTO power of 230 hp. Dividing by 0.87 is the same as multiplying by 1/0.87, or 1.15:

200 divided by 0.87 = 230 rated PTO hp

$$200 \times \frac{1}{0.87} = 200 \times 1.15 = 230 \text{ rated PTO hp}$$

Table 1 shows these ratios for various tractor types and soil conditions.

Matching Implements

When matching a tractor and implement, three important factors must be considered:

- The tractor must not be overloaded, or early failure of components will occur.
- The implement must be pulled at the proper speed, or optimum performance cannot be obtained (Fig. 16).
- The effects of soil conditions on machine performance must be considered. A given tractor has a specific amount of power available. This available power is used for:

a. Moving the tractor over the ground

b. Pulling the implement over the ground

c. Powering the implement for useful work

Fig. 16 — Match Implements to Tractors so Proper Speed Is Maintained — Usually 4.5 mph or More

There are two steps that can be taken to help avoid tractor overloading. First, match the power to the implement for higher field speeds to reduce the loading and wear on the tractor power train. Second, ballast the tractor so you will have from 10% to 15% wheel slip for good tractive conditions. Overloading the tractor with too large an implement and using excessive ballast is a major reason for early power train failures.

While the exact procedure for determining available drawbar horsepower for a specific tractor and a set of conditions is very complicated, we can provide some useful guidelines based on practical field experiences.

Table 2 can be used to estimate the usable drawbar power and the ratio of maximum power to usable drawbar power for various soil conditions.

Soil Conditions — Usable Power		
Condition of Soil	Usable Drawbar Power as a Percentage of Rated PTO Power*	Ratio of Rated PTO Power to Usable Drawbar Power
TWO-WHEEL DRIVE		
Firm	61	1.64
Tilled	57	1.75
Soft/Sandy	47	2.13
FRONT-WHEEL AUXILIARY DRIVE (FWAD)		
Firm	65	1.54
Tilled	62	1.61
Soft/Sandy	55	1.82
FOUR-WHEEL DRIVE		
Firm	66	1.52
Tilled	64	1.56
Soft/Sandy	60	1.67
CRAWLERS		
Firm	70	1.43
Tilled	68	1.47
Soft/Sandy	66	1.52

*If tractor does not have a rated PTO power, assume a rated PTO power of 87% of engine power, or 115% of maximum drawbar power at rated engine speed (as determined on concrete).

Table 2 — Soil Conditions — Usable Power

Table 2 is based on properly ballasted tractors with bias ply tires. If radial ply tires are used, performance would be increased 5% to 10%. Check your operator's manual for recommendations.

The softer or looser the soil conditions, the greater the amount of power that will be consumed because of greater rolling resistance. This reduces the available usable drawbar power (Fig. 17).

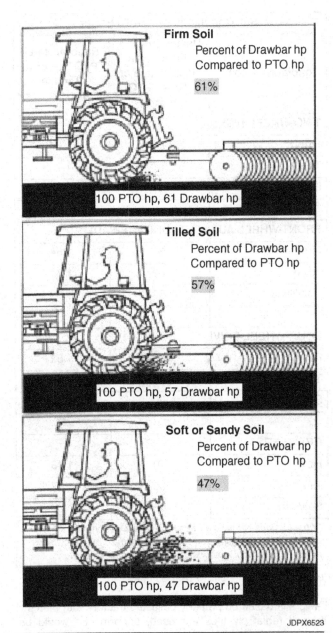

Fig. 17 — As Softer Soil Conditions Are Encountered, Usable Drawbar hp as a Percentage of Rated PTO hp Decreases

It is sometimes helpful to know the ratio of rated PTO hp to usable drawbar hp. The ratio figures in Table 2 make it easier to calculate needed maximum PTO hp when the drawbar hp required by an implement is known.

If a V-ripper is to be used with a two-wheel drive tractor on firm soil, the required usable drawbar power would need to be multiplied by 1.64 to obtain rated PTO power. If the surface has been tilled, the PTO power would need to be 1.75 times the usable drawbar power.

If the implement size is known, the required tractor size can be determined.

Example: If you need 100 usable drawbar horsepower for firm soil, the tractor PTO power would need to be 100 x 1.64 or 164 hp. For a tilled surface, the tractor PTO power would need to be 100 x 1.75 or 175 hp.

The following example shows how to use Table 2. Suppose you want to know how large a tandem disk harrow could be pulled with a front-wheel assist tractor rated at 140 PTO horsepower (Fig. 18). Assume firm soil conditions.

Table 2 shows that an FWAD (front-wheel auxiliary drive) tractor has 65% of rated PTO hp available, or usable, at the drawbar for firm soil conditions. Thus, 65% of 140 PTO hp = 91 hp. From Table 1, let's assume a draft of 340 pounds per foot, or 4.5 hp per foot of width.

91 hp divided by 4.5 = 20.2 ft (6.16 m)

Fig. 18 — Matching a Heavy Disk to a 140 PTO hp Front-Wheel Auxiliary Drive Tractor

How large a four-wheel drive tractor would be needed to pull a 9-shank V-ripper at 4.25 mph (Fig. 19)? Assume a draft of 1,900 pounds per shank (Table 1).

Fig. 19 — The V-Ripper Needs a Four-Wheel Drive Tractor of at Least 331 Engine hp

Table 1 shows 21.5 usable hp per shank needed. Usable hp needed = 9 x 21.5 = 193.5 hp. Assuming firm tractive conditions, Table 2 indicates 66% of rated PTO hp as being usable. Thus, 193.5 hp is 66% of the required PTO hp rating.

193.5 divided by 0.66 = 293 rated PTO hp

Estimating Power Requirements

The four-wheel auxiliary drive tractor would need to be rated at 293 PTO hp or 1.15 x 293 = 337 engine hp. Table 2 indicates engine horsepower as being 115% of maximum PTO hp.

In determining the 87% and 115% conversions from Table 2, an average of several individual tractor tests were considered. These estimates were developed to provide usable approximations as a means of properly matching tractors to implements.

Metric equivalents of these examples are given later in this chapter.

Determining Wheel Slip

Proper wheel slippage is important in order to have the most efficient use of tractor power. Too much slippage wastes energy and reduces available power. If the tractor is overballasted, wheel slippage will be low, but rolling resistance will be excessive, again wasting power. A more serious problem with overballasting is the danger of overloading the power train.

Wheel slip can be determined by tractor instrumentation or the following formula (Fig. 20).

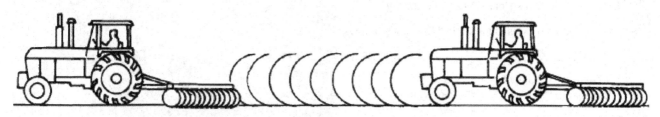

Number of Turns, Loaded = 10

Number of Turns, No Load = 9

Wheel Slip = $\frac{(10 - 9) \times 100}{9}$ = 11%

Fig. 20 — Proper Wheel Slippage Is Important for Efficient Use of Power

First, determine the distance the tractor travels under load in ten wheel revolutions. Next, count the number of turns the wheel makes without the load to travel the same distance.

Wheel slip, percent =

$\frac{\text{(No. of turns, loaded} - \text{No. of turns, no load)} \times 100}{\text{No. of turns, no load}}$

Example:

Number of wheel revolutions, loaded = 10

Number of wheel revolutions, no load = 9

Wheel slip = $\frac{(10 - 9) \times 100}{9}$ = 11%

Wheel slippage should be approximately 8% to 12% for tractors operating on firm soil under a heavy load.

Sizing for Critical Work

In Chapter 4 we discussed matching machine size to fit time available. An important management task is to properly match power units to these machines.

Let's take an example using a typical Corn Belt situation in which we want to select a chisel plow and tractor. Suppose your records show the following:

SAMPLE RECORD OF WORKING TIME			
Month	Calendar Days	Average Working Days	Average Total Working Hours/Month
October	31	18	96
November	30	15	75
March	31	14	70
April	30	16	96
TOTAL	122	63	337

Table 3 — Sample Record of Working Time

Do not base machinery selection on average weather conditions, because weather conditions will vary from year to year. Suppose that in your farming location a maximum of 170 working hours are available in the years of the poorest weather conditions. This means sizing the tractor and plow for 170 working hours.

Fig. 21 — Example of How to Match Implement Size and Tractor to Available Working Time

Even then, don't count on keeping a tractor and plow operating continuously. Some experts say that, at best, 85% of available working time is spent in the field. This further reduces actual working time to 85% of 170 hours or a total of 144 hours.

For example, if you use a coulter chisel plow on 1,000 acres every year in a medium clay loam with a firm surface, how large a disk-chisel plow and tractor would you need?

To chisel plow 1,000 acres in 144 hours, a capacity of 6.94 acres per hour would be needed. Assuming an 82% field efficiency and a speed of 5.5 miles per hour, the size of the chisel plow would be:

$$\text{EFC} = \frac{\text{Speed} \times \text{Width}}{8.25} \times \frac{\text{Field Efficiency}}{100}$$

$$\text{Width} = \frac{8.25 \times \text{EFC}}{\text{Speed}} \times \frac{100}{\text{EFC}}$$

$$= \frac{8.25 \times 6.94}{5.5} \times \frac{100}{82} = 12.7 \text{ feet}$$

A chisel plow with a width of at least 12.7 feet would be needed.

Table 1 indicates 7.3 drawbar horsepower per foot of width for the chisel plow. To determine usable drawbar horsepower needed, multiply the horsepower per foot by the width.

12.7 x 7.3 = 92.7 usable drawbar horsepower needed

Table 2 shows a ratio of 1.64 to convert usable drawbar power to rated PTO power for a two-wheel drive tractor on firm soil. If the tractor were a front-wheel auxiliary drive tractor, the ratio would be 1.54. Tractor size needed:

Two-wheel drive: 1.64 x 92.7 = 152 rated PTO hp

FWAD: 1.54 x 92.7 = 143 rated PTO hp

Metric Equivalents

The metric unit for power is kilowatt.

Force is measured in newtons (N) or kilonewtons (kN).

1 hp = 0.746 kW; 1 kW = 1.34 hp

1 kN = 224.8 pounds force; 1 N = 0.225 pounds

The formula for kilowatts is:

$$\text{Drawbar Power, kW} = \frac{\text{Draft, kN} \times \text{Speed, km/h}}{3.6}$$

Example: If the draft of a trailing implement, such as a disk harrow, is measured at 11.1 kilonewtons and is pulled at a speed of 8 kilometers per hour, what is the drawbar kilowatt?

$$\text{Drawbar Power, kW} = \frac{11.1 \text{ kN} \times 8 \text{ km/h}}{3.6} = 24.7 \text{ kW}$$

The formula can be rearranged and used to determine speed.

The formula is:

$$\text{Speed, km/h} = \frac{\text{Drawbar Power, kW} \times 3.6}{\text{Draft, kN}}$$

Example: A tractor is pulling a plow with a total draft of 22.2 kilonewtons. How fast can the plow be pulled if the tractor has 50 drawbar kilowatts?

$$\text{Speed} = \frac{50 \text{ kW} \times 3.6}{\text{Draft, kN}} = 8.1 \text{ km/h}$$

The formula to determine draft is:

$$\text{Draft, kN} = \frac{\text{Power, kW} \times 3.6}{\text{Speed, km/h}}$$

Example: Assume a 110 kW tractor operates at a speed of 8.9 km/h while pulling a field cultivator. The draft of the field cultivator is 4.96 kN per meter of width when used in a given field (see Table 16, Appendix).

What width of cultivator could be pulled?

 Step 1: Draft = $\dfrac{110 \text{ kW} \times 3.6}{8.9 \text{ km/h}}$ = 44.5 kN

 Step 2: Width, meters = $\dfrac{\text{Draft}}{\text{Draft per meter}}$ =

$\dfrac{44.5 \text{ kN}}{4.96 \text{ km/h}}$ = 9 meters

We would probably select a field cultivator 8 to 9 meters wide.

Example: A 5-bottom, 46-cm plow operates in medium soil with firm surface conditions. Information from Table 16, Appendix shows 30 drawbar kW per meter of width for medium soil.

 Plow width = 5 × 46 cm × m/100 cm = 2.3 meters

2.3 meters × 30 kW per meter = 69 drawbar kW needed

Example: If tractor size is known, determine how large an implement can be pulled.

Tractor size — 110 PTO kW, two-wheel drive

Speed — 8 km/h

Soil condition — firm for traction (see Table 2)

Draft — 5.83 kN per meter of width

Usable drawbar power kW — 61% of rated PTO

PTO kW = 0.61 × 110 = 67.1 kW

Width = $\dfrac{67.1 \text{ kW} \times 3.6}{8 \text{ km/h} \times 5.83 \text{ kN}}$ = 5.18 meters

Summary

Remember, when matching implements and tractors, a margin of error is needed for variations in field conditions. The tables given in this chapter are intended only as guidelines. Use your own figures where possible.

Test Yourself

Questions

1. How much work is done in lifting a 100-pound weight a distance of 30 inches?

2. What drawbar horsepower is required to pull a disk with a total draft of 4,000 pounds at a speed of 5 mph?

3. How fast could you pull a 10,000 pound load with a tractor that has 100 usable drawbar horsepower?

    a. 3.75 mph

    b. 5.62 mph

    c. 6.75 mph

4. A tractor using 60 drawbar horsepower is pulling an 18-foot tandem disk at a speed of 5 mph. What is the draft per foot of width for the disk?

    a. 375 lb/ft

    b. 250 lb/ft

    c. 2,500 lb/ft

5. A front-wheel auxiliary drive (FWAD) tractor has a rating of 150 PTO hp. What would be the maximum usable hp for each of the three soil conditions:

    a. Firm?

    b. Tilled?

    c. Soft?

6. A chisel plow is to be used in heavy soil conditions where the resistance is 600 pounds per foot of width. Soil condition is considered firm. What size four-wheel drive tractor would be needed to pull the chisel at a speed of 6.0 mph if the chisel is 14 feet wide?

7. You have a two-wheel drive tractor rated at 100 PTO hp and want to buy a moldboard plow. You will plow 8 inches deep with most of the soil being a tilled condition. What size plow would you buy? Assume firm soil conditions for the tractor.

8. You have 400 acres of land to prepare for wheat seeding and must go over the tilled ground twice in 16 working days with a tillage tool that has 300 pounds of soil resistance per foot. Assuming it needs to be pulled at 5 mph, what size tool and FWAD tractor would be needed? For each working day you can average 8 hours in the field, and field efficiency is 80%.

9. Why does it require more horsepower in soft soil to pull the same implement than in firm soil?

II

Estimating Machinery Costs

	Chapter
Estimating Fixed Costs	6
Estimating Fuel and Lubricant Costs	7
Estimating Repair Costs	8
Total Costs for Machines and Operations	9

Estimating Machinery Costs

Knowing how to accurately estimate costs for owning and operating farm machinery is an important step to becoming a first-rate manager. Once you know how to estimate costs, it will be much easier to come up with those profit-making decisions.

This section of Machinery Management provides information on how to estimate fixed costs, operating costs, and total costs for any machine, according to its initial cost and annual use.

This second section, along with Section I, will complete coverage of the basic information needed to make the machinery management decisions that will be discussed in Section III.

Estimating Fixed Costs

Fig. 1 — Know How to Estimate Costs to Be a Better Manager

Introduction

One of the most important costs influencing profit in farming operations is the cost of owning and operating machinery. Machinery costs are one of the few costs that good management can minimize. Learning how to realistically estimate machinery costs will aid in reducing production costs (Fig. 1).

Chapter Objectives

- Identify fixed and operating costs
- Calculate depreciation using straight-line, sum-of-the-digits, and declining-balance methods
- Identify methods to reduce fixed costs

Realistic cost estimates play an important role in every machinery management decision — when to trade, which size to buy, and how much to buy.

There are two main types of machinery costs:

1. Fixed costs depend more on how long a machine is owned, rather than how much it is used (Fig. 2).

2. Operating costs vary in proportion to the amount of machine use.

Fixed costs discussed in this chapter include:
- Depreciation
- Taxes
- Shelter
- Insurance
- Interest

Operating costs, discussed in later chapters, include:
- Fuel
- Lubrication
- Maintenance
- Repairs
- Labor

The distinction between fixed costs and operating costs is clear for all items listed except depreciation and repairs. While depreciation is more a fixed cost than an operating cost, it is somewhat affected by the amount a machine is used, particularly if the annual use is unusually high or low.

On the other hand, repairs usually vary according to amount of use, but the need for some repairs seems to result from deterioration due to the age of a machine, as well as how much it is used.

Consider fixed costs and their effect on machinery management.

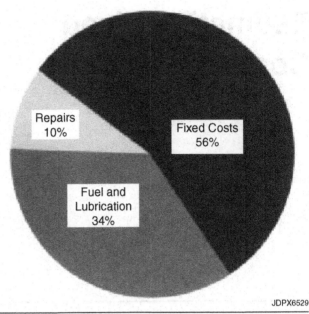

Fig. 2 — Fixed Costs Are Usually the Largest Machinery Cost Items

Depreciation

Depreciation is the loss in value of a machine due to time and use. Often, it is the largest of all costs.

Machines depreciate, or decrease in value, for several reasons, including:

- Age — Even though model changes may have resulted in little difference in the function of a machine, a newer machine is worth more than an old one.

- Wear — As use of a machine increases, wear increases and reliability is reduced, while the need for repairs increases.

- Obsolescence — If there has been a major model change or if a machine no longer has enough capacity, its value may be greatly reduced, even though it may not be worn out. New machine concepts may also be introduced, which may make existing similar machines obsolete (Fig. 3).

Fig. 3 — New Machines May Make Current Machines Obsolete

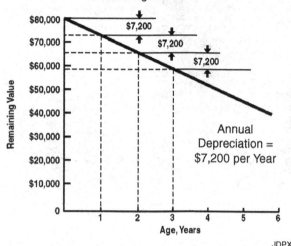

Fig. 4 — The reduction in value is the same each year for straight-line depreciation

Three different ways to calculate depreciation are described below. To learn to estimate depreciation costs, consider these depreciation methods:

- Straight-line depreciation
- Sum-of-the-digits depreciation
- Declining-balance depreciation

Straight-Line Depreciation

The straight-line depreciation method is an equal reduction of value used for each year the machine is owned (Fig. 4). This method can always be used to estimate costs over a specific period of time, provided the proper salvage value is used for the age of the machine.

$$\text{Annual Depreciation (AD)} = \frac{\text{Price} - \text{Salvage Value}}{\text{Years Owned}}$$

The salvage value is a machine's worth at the end of its useful life.

Suppose a self-propelled swather is purchased for $80,000 and its useful life is assumed to be 10 years. Assume its salvage value at 10 years to be 10% of new cost.

$$\text{AD} = \frac{\$80{,}000 - 10\% \text{ of } \$80{,}000}{10}$$

$$= \frac{\$80{,}000 - \$8{,}000}{10} = \$7{,}200 \text{ per year}$$

As shown in Fig. 4, $7,200 per year is the effective rate of depreciation for an $80,000 swather, assuming a straight-line depreciation, a life of 10 years and a salvage value of 10% of new cost.

The straight-line depreciation method is not quite accurate for giving the true value of a machine near the end of its assumed life. In actual practice, machines depreciate much faster in the first few years than in later years.

The best use for the straight-line depreciation method is for estimating costs over the entire life of the machine (Fig. 5). As long as the salvage value of a machine is its actual value at the end of its life, average annual depreciation costs can be estimated with this method.

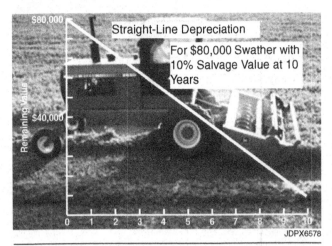

Fig. 5 — Straight-Line Depreciation Applied to an $80,000 Swather

Sum-of-the-Digits Depreciation

The sum-of-the-digits depreciation method is more accurate for estimating the true value of a machine at any age, because the annual depreciation rate decreases as the machine gets older (Fig. 6).

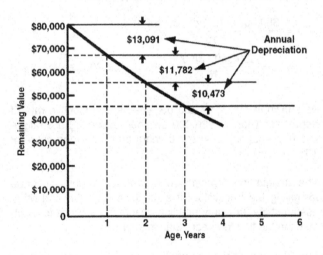

Fig. 6 — With the Sum-of-the-Digits Method, Annual Depreciation Decreases Each Year

The amount of depreciation by the sum-of-the-digits method is determined in three steps:

1. Add up the numbers representing the years covered by the depreciation period.

2. Divide the total depreciation by the sum of the digits of the years for the depreciation period.

3. Proportion the depreciation in reverse of the years over which the depreciation occurs, as shown by the following example.

Applying the sum-of-the-digits method to an $80,000 swather and assuming an $8,000 salvage value after 10 years, the depreciation schedule would be as shown in Fig. 7.

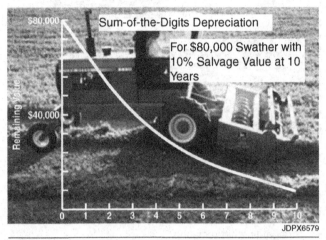

Fig. 7 — Sum-of-the-Digits Depreciation Applied to an $80,000 Swather

The total amount to be depreciated is:

Total value − 10% of total value

$80,000 − $8,000 = $72,000

As in the example for straight-line depreciation, the swather would be depreciated over 10 years. Following the three steps given:

1. The sum-of-the-digits for 10 years is:

10 + 9 + 8 + 7 + 6 + 5 + 4 + 3 + 2 + 1 = 55 depreciation units

2. Total depreciation per unit = $72,000 divided by 55:

$$\frac{\$72,000}{55 \text{ units}} = \$1,309.09 \text{ per unit}$$

3. Each year's depreciation would be calculated as follows for the first 3 years:

First year = 10 × $1,309.09 = $13,090.90

Second year = 9 × $1,309.09 = $11,781.81

Third year = 8 × $1,309.09 = $10,472.72

The tenth year would be the final year of depreciation and would be:

1 × $1,309.09, or $1,309.09

Declining-Balance Depreciation

The declining-balance depreciation method better reflects the actual value of a machine at any age than either the straight-line method or the sum-of-the digits method.

A machine depreciates a different amount for each year with the declining-balance method. However, the percent of depreciation remains the same each year.

For instance, Fig. 8 shows an annual depreciation of 20% of the remaining value. By decreasing the values by 20% each successive year, the declining-balance depreciation method will give a close estimate of the remaining value for tractors or equipment.

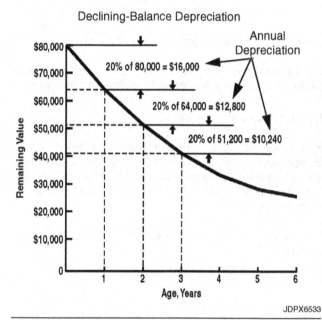

Fig. 8 — The Declining-Balance Method Is Best for Estimating On-the-Farm Remaining Values

The value of a machine on the farm is sometimes referred to as the "as is" value. This value assumes the machine must be sold at a farm auction or on the open market without a trade-in.

The formulas for declining-balance depreciation are shown below. For a more practical and quicker approach, tables will be used.

First, consider the principle of a declining-balance formula for an $80,000 swather, assuming a life of 10 years (Fig. 9).

Fig. 9 — Declining-Balance Depreciation Method for an $80,000 Swather

The formula for remaining value is:

$$RV = P \times (1 - \frac{r}{L})^y, \text{ where:}$$

RV = remaining value

P = initial price

r = rate of depreciation compared to the straight-line method

If r = 2, the schedule is referred to as a double-declining-balance method. With r = 2, the remaining value decreases 20% each year.

L = life, years

y = age at which remaining value is determined

Assumptions:

P = $80,000

r = 2.0

L = 10 years

At y = 1, at the end of the first year of ownership the remaining value (RV) is:

$$RV = \$80,000 \times (1 - \frac{2}{10})^1$$
$$= \$80,000 \times (1 - 0.2)^1$$
$$= \$80,000 \times 0.8 = \$64,000$$

If y = 2, or when the machine is 2 years old, the remaining value (RV) is:

$$RV = \$80{,}000 \times (1 - \frac{2}{10})^2$$
$$= \$80{,}000 \times (0.8 \times 0.8) = \$51{,}200$$

Simply, the declining-balance method works on the basis that whatever value a machine has at the beginning of the year, it will be worth a fixed percentage of that value one year later.

In the case of the new $80,000 swather, its value one year later is 80% of $80,000. In the subsequent year, the remaining value will be worth 80% of $64,000, or $51,200. Using this method, the value of the machine never reaches zero.

In the past few years, a careful study of the "as is" values of agricultural machinery has indicated that on-the-farm remaining values more nearly fit the declining-balance method than either the straight-line or sum-of-the-digits method (Fig. 10).

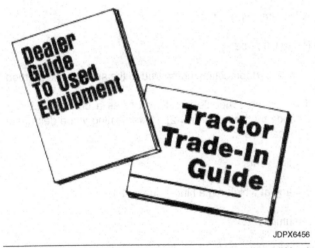

Fig. 10 — *Dealer's Trade-In Guides Provide Reliable Values of Used Equipment*

In actual practice, the first-year depreciation is considerably higher than in later years (Fig. 11). To provide a more accurate method for estimating the value of machines, a first-year correction factor is added to the declining-balance formula.

Fig. 11 — *Farm Machinery Has a High First-Year Depreciation*

For all combines and tractors, including four-wheel drive tractors and crawlers, the remaining-value (RV) formula is:

$RV = \text{list price} \times 0.67 \times 0.94^y$

In this formula, the correction factor for the first-year depreciation is 0.67. The annual depreciation factor is 0.94. The "y" exponent, the age of the machine in years, indicates how many times 0.94, or the annual depreciation factor, is multiplied times itself.

For example, a $50,000 tractor is worth $27,825 after 3 years:

$RV = \text{list price} \times 0.67 \times 0.94^3$

$= \$50{,}000 \times 0.67 \times 0.94^3$

$= \$27{,}825$

For all farm machines other than tractors and combines, the formulas are:

Forage harvesters: $RV = \text{list price} \times 0.67 \times 0.90^y$

All others: $RV = \text{list price} \times 0.67 \times 0.92^y$

Remaining values are listed in Table 1 as percentages of list prices for all tractors and farm machinery. Fig. 12 depicts the remaining values for tractors and combines as a percent of list price.

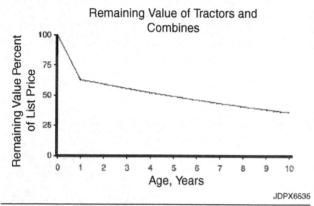

Fig. 12 — Chart Showing Average Remaining Values for Tractors and Combines

To show how the table works, use the following example. Suppose a combine had a list price of $240,000. What would be the remaining value and total depreciation after it is 6 years old?

From Table 1, the remaining value of the combine is 46.22% of list price.

46.22% of $240,000 = 0.4622 x $240,000 = $110,928

Total depreciation is $240,000 − $110,928 = $129,072

Remember that tables and formulas used to estimate "as is" values are based on the average of large numbers of reports from across the country. While they provide a good estimate of typical "as is" values, they cannot possibly take into account used equipment values in times of shortage, local situations, or variations in models. Therefore, use them only for estimates and not to establish an accurate base for a specific trade.

Information in Table 1 can also be used as a guide to estimate the remaining value of equipment that was purchased used. Once used equipment is purchased, remaining values shown in Table 1 for specific age will still apply.

Remaining Value as Percent of List Price			
Age, Years	Tractors & Combines	Forage Harvesters	All Others
0	67.00%	67.00%	67.00%
1	62.98%	60.30%	61.64%
2	59.20%	54.27%	56.71%
3	55.65%	48.84%	52.17%
4	52.31%	43.96%	48.00%
5	49.17%	39.56%	44.16%
6	46.22%	35.61%	40.63%
7	43.45%	32.05%	37.38%
8	40.84%	28.84%	34.39%
9	38.39%	25.96%	31.63%
10	36.09%	23.36%	29.10%
11	33.92%	21.03%	26.78%
12	31.89%	18.92%	24.63%
13	29.97%	17.03%	22.66%
14	28.18%	15.33%	20.85%
15	26.48%	13.79%	19.18%

Table 1 — Remaining Value as Percent of List Price

At first glance there would seem to be a discrepancy in the remaining values when a machine is 1 year old. These values seem low when compared to list prices. The added first year depreciation is there for tax purposes and as such does not reflect a depreciation to the list price after the first year. Remaining values reflect incentive sales by machinery dealers and companies. Table 1 and Fig. 12 are based on list price. Unless otherwise stated, all future cost figures are also based on incentive sales at 85% of list price.

Other Fixed Costs

Now that depreciation has been discussed, we will consider four other fixed costs:

- Taxes
- Shelter
- Insurance
- Interest

The abbreviation for these four costs is TSII.

Taxes

In some states, taxes are paid on machinery in the same manner as for other property. In some cases, a sales tax is also assessed. The annual charge for taxes varies from 1% to 2% of the value of the machine at the beginning of the year.

Shelter

There is a tremendous variation in farmers' use of shelter for agricultural machinery storage. In the dry Southwest, shelter is rarely provided. In colder, more humid areas, nearly all equipment may be stored in some kind of shelter.

Experts agree that if machinery is not stored, more repairs will be needed, machines will deteriorate faster and, in general, higher ownership costs will result.

For this reason, a charge is made for shelter, whether it is actually provided or not (Fig. 13). Typical annual costs for providing shelter, including a service or repair shop, will average 1% to 2% of the remaining value of the machine.

Fig. 14 — Insurance Is Protection Against Loss and Is Part of the Cost of Owning Machinery

Fig. 13 — Shelter for Machinery Is a Fixed Cost

Insurance

Insurance policies are usually carried on more expensive machines, while the risk is usually assumed on the simpler, less expensive machines (Fig. 14). The annual charge for insurance or risk is assumed to be from 0.25% to 0.50% of the remaining value of the machine.

Interest

A large expense item for agricultural machinery is interest (Fig. 15). It is a direct expense item on borrowed capital. Even if cash is paid for purchased machinery, money is tied up that might be available for use elsewhere in the business. Interest rates vary but usually will be in the range of 6% to 10%.

Fig. 15 — Interest Is a Major Fixed-Cost Expense

Estimating Fixed Costs

Taxes, shelter, insurance, and interest (TSII) can be combined in order to estimate costs. If the interest rate is set at 9% and the others combined at 4%, an annual charge of 13% of the remaining value at the beginning of the year would be reasonable for these fixed costs. For such a case, the fixed costs including depreciation and TSII (taxes, shelter, insurance, interest) costs for a tractor with a list price of $100,000 are shown in Table 2.

Total Fixed Costs for a $100,000 Tractor				
		Accumulated Costs		
End of Year	Remaining Value	Depreciation	TSII* Costs	Total Cost
0	85,000			
1	62,980	22,020	11,050	33,070
2	59,201	25,799	19,237	45,036
3	55,649	29,351	26,934	56,284
4	52,310	32,690	34,168	66,858
5	49,172	35,828	40,968	76,797
6	46,221	38,779	47,361	86,139
*TSII = Taxes, Shelter, Insurance, and Interest				

Table 2 — Total Fixed Costs for a $100,000 Tractor

Notice that TSII costs are 13% of $85,000, the estimated value of the tractor at the beginning of the first year. By looking at the total fixed costs (depreciation, taxes, shelter, insurance, and interest), you can see why frequent trading without high annual use results in high per-acre machinery costs.

For purposes of estimates, Table 3 combines all fixed costs. An interest rate of 9% is used in Table 3.

Average Annual Fixed Cost as a Percentage of Original List Price			
End of Year	Tractors & Combines	Forage Harvesters	All Others
1	33.07%	35.75%	34.41%
2	22.52%	24.81%	23.68%
3	18.76%	20.70%	19.75%
4	16.71%	18.33%	17.55%
5	15.36%	16.69%	16.06%
6	14.36%	15.42%	14.93%
7	13.56%	14.39%	14.01%
8	12.90%	13.51%	13.24%
9	12.33%	12.75%	12.57%
10	11.82%	12.07%	11.98%
11	11.37%	11.46%	11.45%
12	10.96%	10.91%	10.96%
13	10.58%	10.41%	10.52%
14	10.23%	9.94%	10.11%
15	9.91%	9.51%	9.72%

Table 3 — Average Annual Fixed Cost as a Percentage of Original List Price

The following example shows how to use Table 3.

A 6-year-old tractor with an annual use of 400 hours originally listed for $80,000. What is the average cost per year for depreciation, taxes, shelter, insurance, and interest after 6 years?

After 6 years, the average annual fixed cost from Table 3 is 14.36% of list price or:

0.1436 x $80,000 = $11,488

The average cost per hour would be:

$11,488 divided by 400, or $28.72 per hour

Low-Annual-Use Fixed Costs

Fixed cost per hour changes according to the amount of annual use (Fig. 16). It is important to understand the total annual fixed cost is not greatly affected by amount of use. Therefore, the more time equipment is used, the lower the average fixed cost per hour or per unit of work.

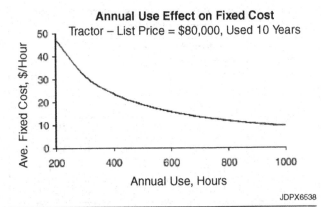

Fig. 16 — Average Annual Fixed Costs per Hour Decrease When Annual Use Increases

To further illustrate how this important principle could affect profit, let's calculate the fixed cost for a self-propelled combine.

Suppose the purchase of a self-propelled combine is being considered and there are two possibilities. One combine lists for $130,000 and the other for $100,000. There are 1,000 acres (405 hectares) of crops to harvest each year. What is the difference in fixed cost per acre, if the combine will be owned for 12 years?

Information in Table 3 shows average annual fixed cost as 10.96%.

For 1,000 acres (405 hectares) of use, the $130,000 combine would have a fixed cost of:

0.1096 x $130,000 = $14,248 per year, or $14.25 per acre

 $14,248/405 hectares = $35.18 per hectare

For 1,000 acres (405 hectares), the $100,000 combine would have a fixed cost of:

0.1096 x $100,000 = $10,960 per year, or $10.96 per acre

 $10,960/405 hectares = $27.06 per hectare

The fixed cost for the larger combine would be $3.29 more per acre for a 12-year ownership. However, expanding the annual use to 1,500 acres (607 hectares) would lower the cost of the $130,000 combine to $9.50 per acre ($23.47 per hectare) (Fig. 17).

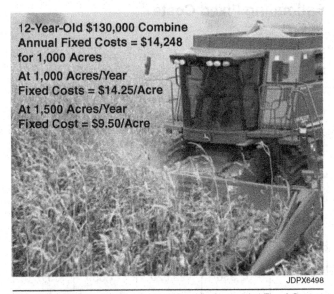

Fig. 17 — The Greater the Annual Use, the Lower the Fixed Cost per Acre (Hectare)

By expanding the area harvested by the larger combine to 1,500 acres, the fixed costs were reduced by $4.75 per acre.

Of course, the $100,000 combine's fixed cost would also be lower. But the question of matching size to the job is critical in this case. The $100,000 combine may not have adequate capacity to get all of the work done within the desired time period.

How to Reduce Fixed Costs

Fixed costs account for 50% to 75% of total machinery costs. They are called fixed costs because they vary only slightly with the amount of annual use. Operating costs, including fuel, lubricants, and labor, vary directly with usage and stay relatively constant on a per-acre basis. Fixed cost per acre can be reduced much easier than operating costs per acre. Reducing fixed costs is even more important when you are in a cost-price squeeze or have cash flow problems. Following are some ways you can reduce fixed costs without lowering production.

- Have the proper amount of equipment
- Lengthen ownership time
- Buy used equipment

Have the Proper Amount of Equipment

Having efficiently matched tractors and equipment will substantially lower fixed costs. For example, consider two 140-horsepower tractors that have an initial cost of $125,000 each. An option is to purchase an 80-horsepower tractor that costs $42,000 to pull some of the lighter draft tools. Owning the tractors for 8 years, results in an average annual fixed cost of 12.9% of list price. See Table 3.

12.9% of the list price of the two 140-horsepower tractors is:

0.129 x 2 x $125,000 = $32,250 per year

Adding the 80-horsepower tractor would have the effect of increasing the fixed costs as follows:

12.9% of $42,000 = $5,418 per year (30% increase)

Certainly the added tractor would make some of the farming operations more efficient, but having equipment that adds to the cost of production without a proportional increase in production should be considered. It is much less expensive to use the larger tractors for light loads than to buy an extra tractor.

Lengthen Ownership Time

Another way to hold down fixed costs of farm machinery is to lengthen ownership. Table 4 illustrates how trading too frequently will increase fixed costs. Equipment listed is a basic set for farming 800 acres of row crops.

LONGER OWNERSHIP LOWERS FIXED COSTS		
Machine	Trade Every 6 Years	Trade Every 10 Years
	Dollars	
$60,000 Tractor	8,614	7,092
$100,000 Tractor	14,360	11,820
$120,000 Combine	17,232	14,184
$75,000 Tillage Tools	11,197	8,985
$25,000 Seeding Equip.	3,732	2,995
Total	55,135	45,076

Table 4 — Longer Ownership Lowers Fixed Costs

The 10-year trading interval lowered the annual fixed costs from $55,135 to $45,076, or $10,059 per year. If the farm size is 800 acres, the added costs for trading every 6 years is $12.57 per acre. Longer ownership of equipment usually requires a good maintenance and repair program. A savings of $10,059 a year would more than meet the needs of a sound maintenance and repair program.

Buy Used Equipment

Buying used equipment is a good way to lower fixed costs, particularly when you are short on finances. There can be a definite financial advantage to purchasing used equipment; however, great care must be taken to be sure that repairs and cost of downtime will not more than offset the savings.

Using depreciation schedules (Table 1) given earlier in this chapter, you can estimate the value of any tractor or piece of equipment of any age.

A better source would be local auction sales. There are several options for purchasing used equipment that include buying from an implement dealer, from an auction sale, or direct from another farmer. Purchasing used equipment from a dealer costs more than the "as is" value. A dealer's price might include a reconditioning charge and/or a limited warranty.

The following examples illustrate the potential for reducing fixed costs by purchasing used equipment.

Example: What would a producer expect to pay for a used $80,000 list price tractor that is 4 years old? How much could this person save in fixed costs?

Table 1 shows that a 4-year-old tractor would have a remaining value of 52.31% of the list price of $80,000.

52.31% of $80,000 is $41,848, which would represent the lower side of what you might expect to pay.

If a tractor is purchased from an implement dealer and a recondition charge is added on, the cost might include the $41,848 plus 25%. This may or may not include a limited warranty.

25% of $41,848 = $10,462

$41,848 + $10,462 = $52,310, the price of the used 4-year-old $80,000 tractor.

Compare the fixed costs for the used tractor and the new tractor. Use Table 3 to estimate costs for the new tractor. After four years, the new $80,000 tractor would have an average annual fixed cost of $13,368 (16.71% of $80,000).

The used tractor would have the following fixed costs for the same 4 years of ownership (Table 5).

Age Years	Remaining Value	Depreciation	TSII	Total
		Dollars		
4	52,310			
5	39,337	12,973	6,800	19,773
6	36,977	2,360	5,114	7,474
7	34,758	2,219	4,807	7,026
8	32,673	2,086	4,519	6,604
	Total Fixed Cost After 4 Years =			40,877

Table 5 — Total Fixed Cost After 4 to 8 Years

The average annual fixed cost of the used tractor would be $40,877 divided by 4, or $10,219 a year. The used tractor would have $13,368 – $10,219, or $3,149 less annual fixed cost. Even if the new tractor is owned for 8 years, it would still have higher annual fixed costs. Try it and see!

This example shows the savings in fixed costs that might be obtained by purchasing used equipment. It is especially economical when you buy equipment without engines and power trains, such as chisels and disks.

The buyer's disadvantage is not knowing the condition of the used machinery or equipment. The potential cost for excessive repairs and downtime is unknown. The subject is explained in greater detail in Chapter 10.

Summary

To a large extent, fixed costs in machinery control profits. An adequate amount of agricultural machinery is necessary to complete all of the important jobs on time. But, it is important to keep fixed cost per acre down by keeping machinery busy.

Fixed cost includes depreciation, taxes, shelter, insurance, and interest.

Test Yourself

Questions

1. List three ways in which machines depreciate.
2. How is straight-line depreciation different from the sum-of-the-digits and declining-balance methods?
3. Using the straight-line method, what is the annual depreciation for a $100,000 combine over a 10-year life if the salvage value is 10% of price?
4. In problem 3, what is the remaining value of the $100,000 combine after 3 years?
5. What would be the average "as is" value for an $80,000 tractor when it is 5 years old?
6. A producer owns a $45,000 swather. Use Table 1 to estimate depreciation during the third year of ownership.
7. What different costs are included in fixed cost?
8. What is the average annual fixed cost for a $60,000 tractor that is 5 years old?
9. Which would have the lowest fixed cost per acre (hectare) of use? A 12-year-old $90,000 combine that is used on 800 acres (324 hectares) per year, or an 8-year-old $70,000 combine used on 800 acres (324 hectares) per year?
10. List three ways to reduce average fixed cost per unit of work done.

Estimating Fuel and Lubricant Costs

Introduction

This chapter covers a subject of great importance in machinery management — estimating fuel and labor costs. Included are methods for estimating these costs for both tractors and self-propelled equipment as well as estimating fuel requirements per acre for various operations.

Fuel and lubricants are true operating costs because fuel consumption is directly proportional to the amount of use (Fig. 1).

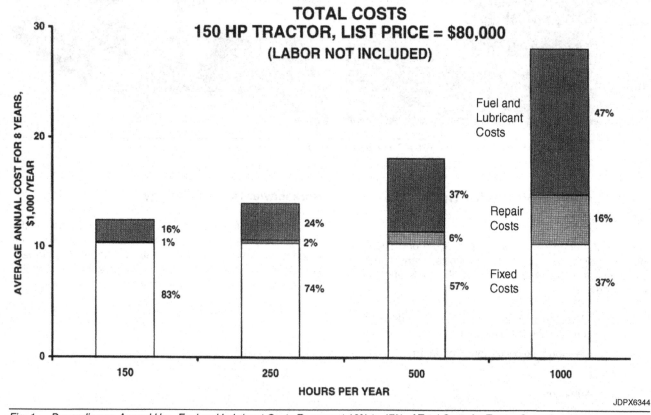

Fig. 1 — Depending on Annual Use, Fuel and Lubricant Costs Represent 16% to 47% of Total Costs for Tractor Operation, Not Including Labor

Chapter Objectives

- Estimate fuel needs
- Estimate average fuel consumption
- Estimate fuel and lubricant costs
- Identify ways to reduce fuel cost

The best way to estimate these costs is to use accurate records on similar machines and operations. However, in cases where such records are not available, it is relatively simple to estimate fuel consumption and lubricant costs for specific operations. Estimating these costs is possible because the amount of fuel consumed is directly related to the amount of energy exerted.

Depending on the type of fuel and the amount of time a tractor or machine is used, fuel and lubricant costs will usually represent at least 16% to over 45% of the total machine costs as shown in the example for a tractor in Table 1.

Also notice from the information in Table 1 that fuel and lubricant costs become a higher percentage of total costs as annual use of the tractor increases. Actually, the fuel and lubricant costs per hour stay constant, but the fixed cost per hour decreases as annual use increases. As a result, operating costs (repairs, fuel, and lubricants) become a higher percentage of annual costs as annual use increases.

Type of Costs[a]	ANNUAL USE, HOURS			
	150	250	500	1000
	Percentage of Total Costs			
Fuel and Lubricants	16.0%	23.8%	36.7%	47.3%
Fixed	83.1%	74.1%	57.1%	36.8%
Repairs	0.8%	2.0%	6.2%	16.0%

a. Labor costs not included.

Table 1 — Tractor Costs

4 Gallons/Hour
(1.5 Liters/Hour)
2 Acres/Hour
(0.8 Hectares/Hour)
4/2 = 2 Gallons/Acre
(15/0.8 = 18.8 Liters/Hectare)

8 Gallons/Hour
(3.0 Liters/Hour)
2 Acres/Hour
(1.6 Hectares/Hour)
8/4 = 2 Gallons/Acre
(30/1.6 = 18.8 Liters/Hectare)

Fig. 2 — The Amount of Fuel Required per Acre Is the Same for Both Plows

Estimating Fuel Needs for Crop Production

The amount of energy needed per acre for performing operations such as disking or plowing is nearly constant, regardless of speed and the size of the tools and tractor being used (Fig. 2).

To better understand how to estimate fuel needs for crop production, consider the following factors:

- Horsepower-hours (kilowatt-hours) of energy
- Fuel types
- Fuel consumption

All three factors are important considerations when estimating fuel needs.

Horsepower-Hours of Energy

The amount of fuel needed per acre is in proportion to the amount of energy required. One way of expressing the amount of energy used is "horsepower-hours" (or kilowatt-hours) (Fig. 3). Reviewing the material presented in Chapter 5 on power ratings will aid in understanding this concept.

Fig. 3 — A 100-hp (75-kW) Tractor Operating at 80% of Maximum Power for 2 Hours Delivers 160 Horsepower-Hours (120 Kilowatt-Hours) of Energy

Horsepower-hours (kilowatt-hours) may be used to measure the work of a tractor in the field. The amounts of energy required for typical farm operations are shown as horsepower-hours per acre units in Table 2. Energy requirements in kilowatt-hours per hectare are shown in Table 17 in the Appendix.

As shown in the previous examples, 1 horsepower delivered for 1 hour is 1 horsepower-hour of energy. Likewise, 1 kilowatt delivered for 1 hour is 1 kilowatt-hour of energy. The type of power used in calculating horsepower-hours (and kilowatt-hours) is PTO power.

Operation	Energy Required, PTO HP-Hrs per Acre	Gallons per Hour		
		Gasoline	Diesel	LP-Gas
Shred Stalks	10.5	1.00	0.72	1.20
Plow 8 Inches Deep	24.4	2.35	1.68	2.82
Heavy Offset Disk	13.8	1.33	0.95	1.60
Chisel Plow	16.0	1.54	1.10	1.85
Tandem Disk, Stalks	6.0	0.63	0.45	0.76
Tandem Disk, Chisel	7.2	0.77	0.55	0.92
Tandem Disk, Plow	9.4	0.91	0.62	1.09
Field Cultivated	8.0	0.84	0.60	1.01
Spring-Tooth Harrow	5.2	0.56	0.40	0.67
Spike-Tooth Harrow	3.4	0.42	0.30	0.50
Rod Weeder	4.0	0.42	0.30	0.50
Sweep Plow	8.7	0.84	0.60	1.01
Cultivate Row Crops	6.0	0.63	0.45	0.76
Rolling Cultivator	3.9	0.49	0.35	0.59
Rotary Hoe	2.8	0.35	0.25	0.42
Anhydrous Applicator	9.4	0.91	0.65	1.09
Planting Row Crops	6.7	0.70	0.50	0.84
No-Till Planter	3.9	0.49	0.35	0.59
Till Plant (With Sweep)	4.5	0.56	0.40	0.67
Grain Drill	4.7	0.49	0.35	0.59
Combine (Small Grains)	11.0	1.40	1.00	1.68
Combine, Beans	12.0	1.54	1.10	1.85
Combine, Corn and Grain Sorghum	17.6	2.24	1.60	2.69
Corn Picker	12.6	1.61	1.15	1.93
Mower (Cutterbar)	3.5	0.49	0.35	0.59
Mower Conditioner	7.2	0.84	0.60	1.01
Swather	6.6	0.77	0.55	0.92
Rake, Single	2.5	0.35	0.25	0.42
Rake, Tandem	1.5	0.21	0.15	0.25
Baler	5.0	0.63	0.45	0.76
Stack Wagon	6.0	0.70	0.50	0.84
Sprayer	1.0	0.14	0.10	0.17
Rotary Mower	9.6	1.12	0.80	1.34
Haul Small Grains	6.0	0.84	0.60	1.01
Grain Drying	84.0	8.40	6.00	10.08
Forage Harvester, Green Forage	12.4	1.33	0.95	1.60
Forage Harvester, Haylage	16.3	1.75	1.25	2.10
Forage Harvester, Corn Silage	46.7	5.04	3.60	6.05
Forage Blower, Green Forage	4.6	0.49	0.35	0.59
Forage Blower, Haylage	3.3	0.35	0.25	0.42
Forage Blower, Corn Silage	18.2	1.96	1.40	2.35
Forage Blower, High-Moisture Ear Corn	5.9	0.63	0.45	0.76
Haul Forage, Corn Silage	4.0	0.42	0.30	0.50

Table 2 — Average Energy and Fuel Requirements

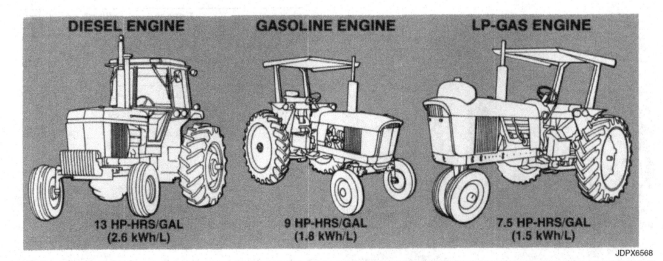

Fig. 4 — Fuel Efficiencies for Three Fuels Used in Farm Tractors

Types of Fuel

Three types of fuel used in farm tractors are diesel, gasoline, and liquefied petroleum (LP-gas), indicated in Table 2. Notice in Table 2 that diesel-engine tractors require fewer gallons per acre than gasoline-engine tractors. LP-gas engine tractors require more gallons per acre than gasoline-engine tractors. Table 17 in the Appendix also gives fuel requirements for farm tractors in liters per hectare.

For the same amount of work, diesel engines will consume about 70% as much fuel as gasoline engines. LP-gas engines consume about 20% more fuel than gasoline engines. Diesel engines are now standard in most tractors, due to their fuel efficiency.

The amount of diesel, gasoline, and LP-gas fuel required per acre for various field operations is also shown in Table 2. These fuel consumption estimates are based on averages generally applying to the corn and wheat belt in the central section of the United States.

For year-round operations of main power units, assume an average fuel efficiency of 9.0 PTO hp-hours per gallon (1.8 kW-h/L) for gasoline, 13.0 PTO hp-hours per gallon (2.6 kW-h/L) for diesel, and 7.5 PTO hp-hours per gallon (1.5 kW-h/L) for LP-gas (Fig. 4).

Table 2 does not include all field operations, and the figures shown may not fit all situations. However, the information can be used to estimate fuel requirements for a series of similar field operations.

Comparing Fuel Consumption of Cropping Systems

	System 1 Conventional Tillage	System 2 Reduced Tillage
Operation	Gallons per Acre	
Disk Stalks	0.45	0.45
Plow	1.68	(Chisel) 1.10
Disk	0.65	—
Pre-Emergence Spray	0.10	0.10
Field Cultivation	0.60	0.60
Plant	0.50	(Till-Plant) 0.40
Cultivate	0.45	0.45
Combine	1.60	1.60
Total	6.68	4.70

Table 3 — Comparing Tillage Methods

The type of cropping system can also be an important factor in fuel costs. Consider the amount of fuel consumed for two different corn productions systems. The tractor and combine used both have diesel engines. Compare the two systems as shown in Table 3. Metric conversions for Table 3 appear in Table 15 in the Appendix.

In this example, the reduced-tillage method saved 1.98 gallons of fuel per acre (18.5 liters per hectare) (Fig. 5). Of course, reducing tillage operations could also reduce yields or increase yields, depending on the crop, soil, weather, and other factors.

Fig. 5 — Reduced-Tillage Methods Conserve Fuel

Estimating Average Fuel Consumption

In order to estimate fuel costs, the average fuel consumption for the machine must be known. While it is possible to estimate fuel requirements for specific levels of draft, it is not worthwhile for purposes of estimating costs.

To better understand how to estimate average fuel consumption, consider these two categories:

- Average fuel consumption for tractors
- Average fuel consumption for self-propelled machines

Estimating average fuel consumption for tractors is the first step for estimating fuel costs.

Estimating Average Fuel Consumption for Tractors

Fig. 6 — On a Year-Round Basis, Farm Tractors Will Operate at Approximately 55% of Rated Power

For most farming operations, a tractor will operate at approximately 55% of its rated power (Fig. 6). Table 4 illustrates the percent of time the tractor operated at each power level.

Rated Power, Percentage	Total Time, Percentage
Over 80%	16.8%
60–80%	23.9%
40–60%	22.6%
20–40%	17.5%
Under 20%	19.2%

Source: Agricultural Engineering Department, University of Illinois

Table 4 — Percentage of Time Tractors Operate at Selected Levels of Power

Because tractors usually operate at an average of 55% of rated power, multipliers that have been determined to estimate fuel requirements. These multipliers are listed in the center column in Table 5. Metric conversions for Table 5 appear in Table 16 in the Appendix. See Metric Equivalents, later in this chapter, for the following calculations in metric measurements.

Engine (Fuel Type)	Average Fuel Consumption (Gallons per Hour per Rate PTO hp)	Typical Pounds per Gallon
Gasoline	0.068	6.1
Diesel	0.044	6.9
LP-Gas	0.080	4.25–4.50

Table 5 — Average Fuel Consumption and Fuel Weights for Different Types of Tractor Engines

Fig. 7 — Diesel-Engine Tractors Average 0.044 Gallons per Hour per Rated PTO HP on a Year-Round Basis

If a diesel-engine tractor has 75 rated PTO horsepower, its fuel use would probably average 3.30 gallons per hour (75 hp x 0.044) on a year-round basis (Fig. 7). A gasoline model of the same size would average 5.1 gallons per hour, and an LP-gas model, 6.0 gallons per hour.

Naturally, some tractors will be more efficient than others. These average figures will not give a perfect answer for every case. However, the variation of any tractor from the average figure will make little difference in estimating costs.

To estimate fuel costs, multiply the average fuel consumption of the tractor by the fuel cost per gallon.

Suppose the average fuel cost for a 60-horsepower, gasoline-engine tractor is to be calculated with the cost of gasoline at $4.00 per gallon. As shown in Table 5, the fuel consumption multiplier for gasoline is 0.068. The answer is derived as follows:

60 hp x 0.068 = 4.08 gallons per hour, the average fuel consumption (Fig. 8).

4.08 gallons/hour x $4.00/gallon = $16.32/hour, the average fuel cost.

Estimating fuel costs is discussed in more detail later in this chapter. For metric formulas, use Table 16 and Table 17 in the Appendix. (Also see Metric Equivalents later in this chapter)

Fig. 8 — Multiply Rated PTO HP by 0.068 to Get Average Fuel Consumption for Gasoline-Engine Tractors

Estimating Fuel Consumption for Self-Propelled Machines

Calculating fuel consumption for self-propelled machines is accomplished the same way as calculating fuel consumption for tractors, except we use average fuel consumption per acre, as shown in Table 2. Sometimes fuel consumption is based on the power of the engine of a self-propelled machine. For our purposes, it will be simpler to use Table 2.

One major difference in estimating fuel consumption is that operations involving self-propelled machines, such as combines or self-propelled forage harvesters, usually involve other machines for transport and storage.

Consider an example of a forage harvesting operation. The total operation consists of chopping the forage, hauling the material to the silo, and using a blower to fill the silo.

Considering the distance from the field to the silo, two tractors and wagons will keep the forage harvester and blower operating almost continuously. Going to the information in Table 2 (or Appendix, Table 17), the following average fuel consumption figures are given:

- Chopping forage (self-propelled harvester, diesel engine) = 3.6 gallons per acre (33.6 L/ha)
- Tractor-powered blower (diesel engine) = 1.4 gallons per acre (13.1 L/ha)
- Transport (two diesel-engine tractors) = 0.3 gallons per acre (2.8 L/ha)

By adding the amount of fuel involved in each operation, a total of 5.3 gallons per acre (49.5 L/ha) is determined as the total amount of fuel needed for the entire forage harvesting operation.

Estimating Average Fuel and Lubricant Costs

After estimating fuel consumption for tractors and self-propelled machines, the next step is estimating average fuel and lubricant cost. This important aspect of machinery management will be discussed in two parts:

- Estimating lubricant costs
- Estimating fuel and lubricant costs

Arbitrary figures for fuel cost per gallon (liter) are used in this book. Always use the actual cost of fuel for each situation as fuel prices vary.

Estimating Lubricant Costs

Fig. 9 — Lubricant Costs Are Approximately 10% of Fuel Costs

Modern tractors and self-propelled machines use a variety of lubricants — engine oil, grease, transmission oil, and hydraulic fluid. The Agriculture Census indicates that lubricant costs are approximately 10% of the fuel cost of agricultural machinery (Fig. 9). Lubricant costs can be estimated simply by multiplying the fuel cost by 10%.

Estimating Both Fuel and Lubricant Costs

Fig. 10 — Keeping Records of Fuel Consumption Is the Best Answer for Accurate Fuel Consumption Estimates

For the most accurate estimate of fuel and lubricant costs, keep records on each tractor or machine (Fig. 10). But if accurate fuel consumption records are not available, suitable cost estimates can be made by using the methods described in this chapter.

Table 2 lists figures useful for estimating fuel and lubricant costs for different field operations. Lubricant costs for various operations including self-propelled machines will average approximately 10% of fuel costs.

Consider a 6-row, 30-inch corn combine with a diesel engine. Calculate the fuel and lubricant costs with diesel fuel costing $4.00 per gallon. (See Metric Equivalents, later in this chapter, to calculate fuel and lubricant costs in metric measurements.)

Information in Table 2 indicates that the fuel consumption for a combine would average 1.6 gallons per acre. First, calculate the fuel cost per acre:

1.6 gal/acre x $4.00/gal = $6.40/acre

At 3.5 acres per hour, the resulting fuel cost for the combine in this example would be:

$6.40/acre x 3.5 acres/hr = $22.40/hr

Because lubricant costs usually average 10% of fuel cost:

$22.40/hr x 0.10 = $2.24/hr lubricant cost

The total fuel and lubricant cost in this situation would be $24.64 per hour ($22.40 + $2.24).

Fuel-Saving Tips

Operating cost will be reduced by cutting fuel consumption whenever possible. Some ideas to consider are:

1. Reduce tillage, when possible. Additional tillage means additional use of fuel (Fig. 11).

Fig. 11 — Each Tillage Trip Eliminated Saves One-Half to Two-Thirds of a Gallon of Fuel per Acre

2. Combine operations and operate the tractor at full load (Fig. 12). This can save one-half gallon or more per acre because a tractor is more efficient when operating at full load. Two trips with a partial load require considerably more fuel than one trip fully loaded using combined operations.

Fig. 12 — Combining Trips Over the Field Can Save One-Half Gallon of Fuel per Acre or More

3. Shift to a higher gear and throttle back when pulling a light load (Fig. 13). It takes one-third more fuel when pulling a light load with the governor control wide open than to gear up and throttle back.

Fig. 13 — Throttle Back and Shift to a Higher Gear for Light-Draft Implements

 CAUTION: Be careful not to overload the tractor with a reduced engine speed and a higher gear.

4. Select tire size and match ballast for proper wheel slippage. For most operations, 10% to 15% wheel slippage gives the best combination of maximum power at the least fuel usage. Too much slippage can waste as much as 2 gallons per acre.

5. Keep the tractor in top running condition. Follow a rigid maintenance schedule and have the dealer give the tractor or machine a periodic service check. A tractor that is operating poorly can waste up to 25% of the fuel.

6. Follow recommended storage practices for fuel tanks. Fuel suppliers can provide some good suggestions on storage. Use rustproof tanks or keep them shaded and use a pressure cap that conforms to local regulations (Fig. 14). Diesel fuel is less likely to evaporate than gasoline. For both diesel fuel and gasoline, leakage and contamination by dirt and water are possible. Fuel tank filters protect engines by removing contaminants.

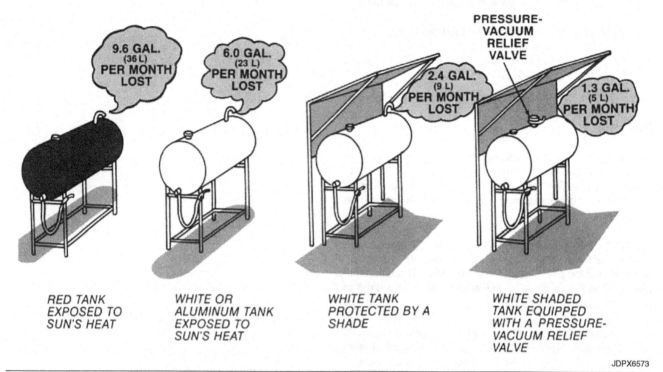

Fig. 14 — Summer Evaporation Losses From a 300-Gallon (1,135 L) Gasoline Storage Tank

Metric Equivalents

As discussed earlier in the chapter, the average annual fuel consumption for diesel tractors is 0.044 gallon per rated PTO horsepower-hour. The metric equivalent can be determined as follows:

1 hp = 0.746 kW; 1kW = 1.34 hp

1 gallon = 3.785 liters; 1 liter = 0.264 gallon

Using the unit-factor method, the fuel consumption in liters per kWh is determined as follows:

$$\frac{0.044 \text{ gallons}}{\text{hp} \times \text{hr}} \times \frac{3.785 \text{ liters}}{\text{gallons}} \times \frac{1 \text{ hp}}{0.746 \text{ kW}}$$

0.223 liters per kW-h

Also see Table 18, Appendix.

The 150 PTO hp tractor example earlier in this chapter would have the following metric equivalents:

150 hp x 0.746 kW/hp = 111.9 kW

111.9 kW x 0.223 liters/kWh = 24.95 liters/h

24.95 liters/h x 0.264 gallons/liter = 6.6 gallons/h

Calculating Fuel and Lubricant Costs in Metric Units

For this example, we will use a four-wheel drive tractor with 350 rated engine horsepower.

As shown earlier, the tractor would have 350 x 0.87, or 304.5 rated PTO hp.

304.5 hp x 0.746 kW/hp = 227 kW

227 kW x 0.223 liters/kWh = 50.6 liters/h

At $1.06 per liter, the fuel cost would be:

$1.06/liter x 50.6 liters/h = $53.63 per hour

Adding 10% of fuel cost for lubricant cost gives the following:

$53.63 + (0.10 x $53.63) = $58.99 per hour

For this example, use a 6-row, 75-centimeter corn combine with a diesel engine. Calculate the fuel and lubricant costs with diesel fuel costing $1.06 per liter.

Information in Appendix Table 17 indicates the combine would have an average fuel consumption of 15 liters per hectare. First, calculate the fuel cost per hectare:

15 liters/ha x $1.06/liter = $15.90/ha

Because lubricant costs usually average 10% of fuel cost:

$15.90 per ha x 0.10 = $1.59 per ha lubricant cost

The total fuel and lubricant cost in this situation would be $17.49 per ha ($15.90 + $1.59).

Summary

Fuel and lubricant costs usually represent 20% to 30% of total machine cost in farming operations. As part of operating costs, fuel and lubricant costs become a larger percentage of total machinery cost as annual use of the machine increases.

For estimating purposes, the amount of energy needed per acre for each type of operation is nearly constant, regardless of the size of the power unit or machine used. The amount of fuel used is proportional to the amount of energy used in performing each field operation.

Fuel requirements for each type of engine — diesel, gasoline, and LP-gas — vary for the same amount of work because of the different fuel efficiencies of each type of engine.

Estimating the average fuel consumption for a tractor or self-propelled machine is the first step in estimating fuel costs for various operations. Lubricant costs usually can be estimated as 10% of fuel cost for both tractors and self-propelled machines.

An important machinery management skill is conserving fuel whenever possible. Fuel prices vary around the country, but can represent important savings in operating costs for the efficient manager.

Test Yourself

Questions

1. Match each type of fuel with its typical weight per gallon:

   a. Gasoline   1.) 6.9 pounds per gallon

   b. Diesel   2.) 4.25–4.50 pounds per gallon

   c. LP-gas   3.) 6.1 pounds per gallon

2. On the average, for year-round operation, farm tractors will operate at:

   a. 75% of maximum power

   b. 55% of maximum power

   c. 45% of maximum power

3. What is the average hourly and annual fuel consumption for a 100 hp (rated PTO) tractor, if it is used 450 hours a year? Also, estimate hourly and annual fuel and lubricant costs with diesel fuel at $4.00 a gallon.

4. (T/F) Fuel and lubricant costs are the largest single cost in owning and operating a tractor.

5. If fuel costs average $5.50 per hour for your tractor, how much would you expect to spend for lubricants?

   a. $1.50 per hour

   b. $2.15 per hour

   c. $0.55 per hour

   d. $8.25 per hour

6. Which type of tractor engine is the most efficient in converting fuel to useful work?

7. How many gallons of diesel fuel would you expect to use for disking, chisel plowing, field cultivating, planting, and combining 1 acre of corn?

8. How much would you save in diesel fuel costs if you were pulling a planter with a 200 rated PTO horsepower tractor and you use part throttle and a higher gear? You are planting 1,000 acres at 10 acres per hour. Assume diesel fuel at $4.00 a gallon and full throttle fuel consumption at 6.5 gallons per hour.

9. Estimate the average fuel and lubricant costs for a 275 rated PTO horsepower tractor with diesel fuel costing $4.00 per gallon.

10. If fuel and lubricant costs are $44.00 an hour, what is the rated engine kW rating of the tractor? Diesel fuel costs $4.00 a gallon. (Hint: Engine kW is 1.15 times PTO kW.)

    a. 227 kW

    b. 195 kW

    c. 261 kW

    d. 145 kW

11. If a diesel tractor uses 15 gallons an hour, how much would it use if it had a gasoline engine of the same size?

12. List three ways to save tractor fuel.

13. Explain why fuel, lubricant, and repair costs become a larger percentage of total cost as annual use increases.

Estimating Fuel and Lubricant Costs

Estimating Repair Costs

Introduction

Fig. 1 — It Is Important to Keep High-Value Operations Going at Full Capacity

Repair costs, usually considered an operating cost, are another important part of machinery costs (Fig. 1). The more a machine is used, the greater is its need for repairs.

Chapter Objectives

- Identify types of repairs
- Calculate lost time from repairs
- Establishing life of equipment
- Estimating repair cost

Fig. 2 — Consider All Repair Costs as an Operating Cost

Some machine components have surfaces that rust, rot, or otherwise deteriorate over the years. Repair costs caused by deterioration, though not necessarily affected by the amount of use, are still considered an operating cost (Fig. 2).

Repairs are necessary for keeping a machine running, as well as for keeping it operating properly. The purpose for repairing a machine is to maintain its reliability and to keep it performing its task properly.

The longer machinery is used, the more repairs are needed to maintain reliability. Reliability expresses the amount of confidence in a machine to perform without an unplanned time loss due to a breakdowns (Fig. 3).

Fig. 3 — Downtime Can Be Expensive

The greater productivity from larger new tractors makes reliability more important to avoid breakdowns or make quick repairs. Even one hour during periods of critical operations in farming is valuable. For example, a breakdown that causes a planting delay of one hour can mean $200 or more in yield loss (Fig. 4).

Fig. 4 — Each One-Hour Delay in Planting Can Result in a Timeliness Cost of $200 or More in Yield Reduction

Yield-loss cost is probably more important than direct cost of repairs. Make it a goal, once a machine is purchased, to provide necessary maintenance and repairs to keep it running at a high level of reliability.

The aspects of repair costs and reliability considered in this chapter include:

- Types of repairs
- Calculating lost-time costs
- Establishing life of equipment
- Estimating repair costs

Types of Repairs

With any machine, there are four types of repairs. These are:
- Routine wear
- Accidental breakage or damage
- Repairs due to operator neglect
- Routine overhauls

Let's compare each of the four types of repairs and compare the effect management might have on either cost or reliability.

Routine Wear

Typical examples of routine wear (or replacement) would include plowshares, disk blades, sickles, tires, and batteries (Fig. 5). Even with the best of care, replacement will be necessary sooner or later.

Fig. 5 — Some Repairs Are Due to Routine or Normal Wear

If a rigid maintenance schedule is followed and the equipment is not abused, an increase of 50% to 100% in parts life can result, depending on the nature of the part.

Except for excessive abuse, the life of plowshares, disk blades, and other soil-engaging parts is usually determined by the soil characteristics. Other parts, like tires and batteries, have a life that is affected by maintenance.

Accidental Breakage or Damage

Machinery accidents can happen, even with the best of operators. Carelessness or rushing a job is more likely to cause costly accidents (Fig. 6). Unfortunately, these types of accidents often involve a frame, axle, housing, or some other part costly to replace. Additionally, the total damage to the machine might be quite extensive.

Fig. 6 — This ROPS Enclosure Will Have to Be Replaced and the Tractor Repaired Due to an Accident

With good management, few, if any, repairs of this nature are needed. Using good judgment can eliminate most accidental breakage.

Repairs Due to Operator Neglect

Occasionally, the time is not taken to perform needed maintenance or minor repairs. Neglected maintenance and minor repairs lead to more serious problems (Fig. 7).

Fig. 7 — Neglecting Maintenance Can Lead to Expensive Overhauls

Putting off maintenance and minor repairs can be costly. Use a good off-season repair program, together with a rigid daily inspection and service schedule, to help avoid expensive repairs (Fig. 8).

Fig. 8 — Periodic Checking Is Needed to Prolong Tire Life

Routine Overhauls

Machines are overhauled to replace worn or defective parts and restore original performance. Overloading and poor maintenance practices can accelerate these repairs by 100% or more. Records of repair costs compiled by the Agricultural Engineering Department, University of Illinois, show that with good management, repair costs can be reduced by 25% or more (Fig. 9). Poor management can raise repair costs 25% above the average.

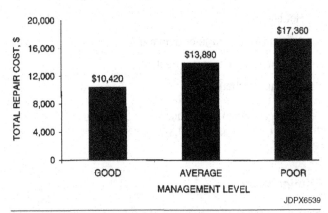

Fig. 9 — Good Management Means Lower Repair Costs

Calculating Lost-Time Costs

Applying simple arithmetic to machine reliability serves as a strong motivation to improve machine operation.

Suppose a combine can harvest soybeans with a 6-row, 30-inch (75-centimeter) header at 3.25 miles per hour (5.2 kilometers per hour). The theoretical capacity is 5.9 acres (2.4 hectares) per hour or 47.2 acres (19.2 hectares) in 8 hours.

Records show that 75% field efficiency is possible in this case with proper management and no breakdowns. At 75% field efficiency, it is possible to harvest 4.4 acres (1.8 hectares) per hour and 35.2 acres (14.25 hectares) in 8 hours.

If just one hour a day is lost due to unplanned stops for minor repairs, capacity would be further reduced to 30.8 acres (12.5 hectares) per day. At a custom rate of $25.00 per acre ($61.75 per hectare), the lost production of 4.4 acres (1.8 hectares) per day would cost an equivalent of about $110.00.

As a general rule, it pays to spend one or two days in the slack season to avoid a one-hour loss when the machine is needed. With increasing need for larger-capacity equipment, every effort should be made to keep machines in top shape (Fig. 10). An excellent maintenance program to keep machines in good operating condition is one of the best possible investments.

Fig. 10 — Reliability Is Even More Important With Larger, High-Capacity Machines

Establishing Life of Equipment

Usually, repair costs are related to how much a machine is used. Therefore, before estimating how much to expect in the way of repair costs, it is necessary to set some limits on the total maximum life a machine can be expected to continue operating. Then we can develop some guidelines for repair cost estimates.

Information in Table 1 indicates the average maximum life to expect from typical farm machinery with average annual usage. It is quite possible to have a tractor or machine in good running condition well beyond the limits shown. To exceed these limits, extra maintenance and repairs would be necessary in most cases.

Machine	Life, Hours
All Wheel-Type Tractors	12,000
Crawlers	16,000
Self-Propelled Combines	3,000
Self-Propelled Windrowers	3,000
Cotton Harvesters	5,000
Tillage Equipment, Mowers	2,000
Planters, Drills, Large Round Balers	1,500
Small and Large Square Balers, Rakes	2,500
Self-Propelled Forage Harvesters	4,000

Table 1 — Estimated Mechanical Life

Estimating Repair Costs

Fig. 11 — Repair Costs Vary in Different Parts of the Country

Repair costs consist of all expenditures for parts and labor for repairs made in a shop or on the farm. In the case of older equipment, an estimate of deferred repair costs should be included. It is difficult to accurately predict repair costs for a specific machine. Repair costs will vary from one geographical section of the country to another because of differences in soils, crops, climate, and operators (Fig. 11).

Machine	1/4 Accumulated		1/2 Accumulated		3/4 Accumulated		Full Life Accumulated		RF1	RF2
	Hours	Cost	Hours	Cost	Hours	Cost	Hours	Cost		
All Wheel Tractors	3,000	6.2%	6,000	25.0%	9,000	56.2%	12,000	100%	0.006944	2.0
Crawlers	4,000	5.0%	8,000	20.0%	12,000	45.0%	16,000	80%	0.003125	2.0
Self-Propelled Combines	750	2.2%	1,500	9.3%	2,250	21.9%	3,000	40%	0.039820	2.1
Cotton Harvesters	1,250	9.9%	2,500	24.4%	3,750	41.3%	5,000	60%	0.074044	1.3
Planters, Drills	375	4.1%	750	17.5%	1,125	41.0%	1,500	75%	0.320000	2.1
Moldboard Plows	500	8.3%	1,000	28.7%	1,500	59.6%	2,000	100%	0.287300	1.8
Chisel Plows, Mulch Tillers, Cultivators, Disk Harrows, etc.	500	10.1%	1,000	26.5%	1,500	46.8%	2,000	70%	0.265240	1.4
Mowers	500	14.2%	1,000	46.2%	1,500	92.0%	2,000	150%	0.461700	1.7
Large and Small Square Balers	625	6.2%	1,250	21.5%	1,875	44.7%	2,500	75%	0.144100	1.8
Large Round Balers	375	7.4%	750	25.9%	1,125	53.6%	1,500	90%	0.434000	1.8
Self-Propelled Forage Harvesters	1,000	3.1%	2,000	12.5%	3,000	28.1%	4,000	50%	0.031250	2.0
Self-Propelled Windrower	750	3.4%	1,500	13.7%	2,250	30.9%	3,000	55%	0.061100	2.0
Rakes	625	8.6%	1,250	22.7%	1,875	40.1%	2,500	60%	0.166300	1.4

Source: American Society of Agricultural Engineers

Accumulated repair costs = list price x RF1 x (total hours/1,000)^{RF2}

Repair cost estimates in this table do not include the effect of inflation.

Table 2 — Accumulated Repair Costs as a Percentage of List Price

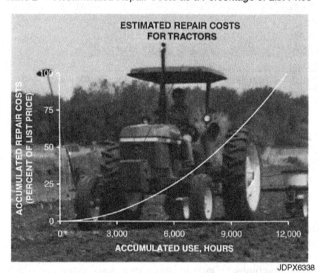

Fig. 12 — Estimated Repair Costs for Wheel-Type Tractors

Fig. 13 — Estimated Repair Costs for Combines

Fig. 14 — Estimated Repair Costs for Cotton Harvesters

Fig. 15 — Estimated Repair Costs for Planters and Drills

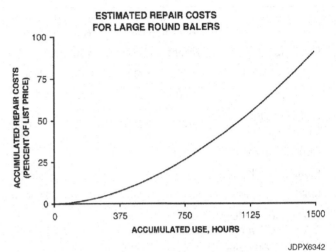

Fig. 16 — Estimated Repair Costs for Large Round Balers

Fig. 17 — Estimated Repair Costs for Disks, Chisels, Plows, and Field Cultivators

Repair costs can be estimated for any machine with the following formula:

TAR = Total Accumulated Repairs

TAR = List Price x RF1 x (hours/1,000)^{RF2}

Table 2 gives a listing of values for RF1 and RF2. Also, listed in Table 2 are estimated repair costs for different amounts of machine use. (See the Appendix for RF1 and RF2 values for other machines.) Data in this table are based on 1998 Standards of the American Society of Agricultural Engineers. By plotting the formula it can be seen that repair costs increase with machine use at an increasing rate (Fig. 13 through Fig. 17).

Example: What are the predicted accumulated repair costs for a $90,000 tractor after 4,000 hours of use?

TAR = $90,000 × 0.006944 × (4,000/1,000)^2

TAR = $10,000

Table 2 is a summary of repair costs for some of the more commonly used machines.

The repair cost formula should not be used when the total hours exceed the expected life.

Information in Table 2 can be used in two meaningful ways:

- To help estimate total costs of a specific machine.
- To compare known repair costs against what is considered average.

The table can also be used to compare estimated and actual repair costs.

Suppose a $50,000 list price tractor was purchased 5 years ago. Assume that $2,790 was spent over that time for repairs, which included all service shop costs, parts installed, and labor. The tractor hour meter shows 2,876 hours. Now, let's compare these repair costs to the estimate in Table 2. According to Table 2, a total of 2,876 hours is nearly one-fourth life. The table shows repair costs average 6.2% at that point.

0.062 × $50,000 = $3,100

In our example, $2,790 was spent, which is $310 less than estimated from Table 2. Unless unusual repair costs are incurred in the next 124 hours of use, the tractor will have close to average repair costs or just under that amount.

Cost per hour = $\frac{\$2,790}{2,876 \text{ hrs}}$ = $0.97 per hour

The estimate from the table is:

$\frac{\$3,100}{3,000 \text{ hrs}}$ = $1.03 per hour

Information in Table 2 can also be used for estimating repair costs for a specific machine. Suppose it is planned to do custom work with a $100,000 self-propelled combine and an estimate of repair costs is needed. Assume the combine will be used 1,500 hours, one-half its mechanical life. The combine operates at 7 acres per hour.

Information in Table 2 shows accumulated repair costs to be 9.3% of the new cost at 1,500 hours. Since 9.3% of $100,00 is $9,300, or $6.20 per hour, repair costs would be $0.89 per acre ($2.19 per hectare) at 7 acres (2.83 hectares) per hour.

Summary

Repair costs can usually be classified as an operating cost. Repairs are an important part of machinery management, because they are necessary to maintain a machine's reliability and keep it performing at maximum capacity.

Reliability expresses the amount of confidence placed in a machine to perform without unplanned time losses.

It is more important than ever to avoid or to be able to quickly repair machine breakdowns. Because of the larger machine size on farms today and the resulting increased productivity, good management of repairs and the resulting increase in reliability is well justified.

There are four main types of repairs:

- Routine wear
- Accidental breakage or damage
- Repairs due to operator neglect
- Routine overhauls

It pays to get machines ready ahead of time. Make all necessary repairs and have machines adjusted and ready to go before they are needed.

Estimating Repair Costs

Test Yourself

Questions

1. What are the four reasons for repairs?

2. What is the main purpose of repairs?

3. Define reliability.

4. Calculate the cost of repairs during the first 3,500 hours of use for a $90,000 tractor.

5. If a cotton harvester costs $100,000 and harvests 5,000 bales in 2,500 hours, what is the average repair cost per bale?

    a. $2.67

    b. $4.88

    c. $6.62

    d. $7.16

    e. $10.24

Total Costs for Machines and Operations

9

Introduction

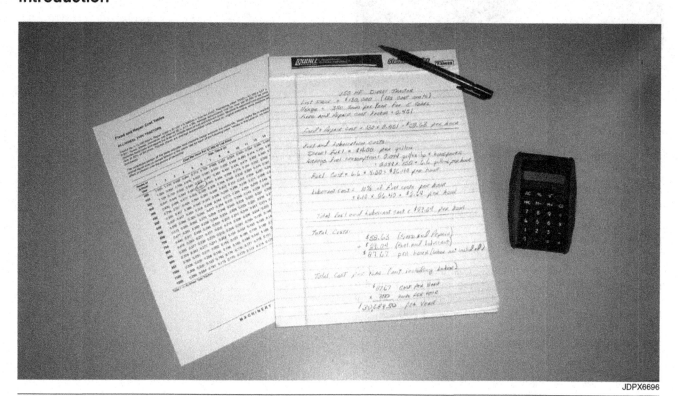

Fig. 1 — To Make Important Management Decisions, Know How to Estimate Total Costs

In the past three chapters, methods for estimating fixed costs, operating costs and repair costs were presented. These three costs will now be combined in this chapter to estimate total cost. Accurate total-cost estimates are the basis for important management decisions and will be used throughout the following chapters (Fig. 1).

This chapter will provide information on how to estimate total cost for the use of an individual machine, and for operations involving a combination of two or more machines. Costs for entire machine systems can be estimated once the procedure for estimating costs for specific machines is mastered.

Chapter Objective

- Estimating total machine cost

Cost for Individual Machines

Some of the tables in the Appendix have been developed for making cost estimates. These tables are intended to be used only as one source of cost estimates. Your own accurate records (Fig. 2) would be a better source than these tables, except for when you are thinking of buying a new type of machine. The cost tables in the Appendix (Table 1 through Table 13) are a combination of information in Chapters 6, 7, and 8, except that fuel, lubricants, and labor costs are not included and must be figured separately. These tables include fixed costs and repair costs.

Fig. 2 — Accurate Records Provide the Best Information for Cost Estimates

Using Machine Cost Tables

To illustrate how to use the machine cost tables, look at the tractor costs shown in Table 1 (same as Table 1 of the Appendix). Costs will be calculated for a 150-hp diesel tractor with a list price of $130,000. It is used 350 hours a year for 5 years, and has a diesel engine with fuel costs of $4.00 per gallon ($1.06 per liter).

ALL WHEEL-TYPE TRACTORS

Consider this situation:

- Useful life of a tractor = 12,000 hours (while the tractor has a useful life of 12,000 hours, the farmer plans to use it only 350 hours per year)

- Repair costs = $1,000 \times 0.006944 \times (hours/1,000)^2$

- Remaining value, RV ($) = $1,000 \times 0.67 \times 0.94^y$

- Taxes, shelter, insurance, and interest ($) = $0.13 \times RV$

- Average accumulated cost per hour per $1,000 of list price (includes fixed costs and repairs but does not include labor or fuel and lubricant costs)

- Machine is assumed to be purchased new at 85% of list price. Figures shown are average accumulated costs based on original ownership since machine was purchased.

The underlined portion of the table indicates where total machine usage exceeds the useful life. When useful life is reached, the annual repair cost is carried forward at the same rate for each of the following years. Values in this table do not include the effect of inflation over the period of ownership. Use the tables in the Appendix to estimate costs for specific situations.

Cost per Hour per $1,000 of List Price

Hours of Annual Use	\multicolumn{15}{c}{Age, Years (y)}
	1	2	3	4	5	6	7	8	9	10	11	12	13	14	15
200	1.655	1.129	0.942	0.841	0.775	0.726	0.688	0.656	0.629	0.605	0.584	0.565	0.547	0.531	0.516
250	1.325	0.904	0.756	0.676	0.623	0.585	0.555	0.530	0.509	0.490	0.474	0.459	0.446	0.434	0.422
300	1.104	0.755	0.632	0.565	0.522	0.491	0.467	0.447	0.430	0.415	0.402	0.390	0.380	0.370	0.362
350	0.947	0.648	0.543	0.487	0.451	0.425	0.404	0.388	0.374	0.362	0.352	0.342	0.334	0.326	0.320
400	0.830	0.569	0.477	0.429	0.398	0.376	0.358	0.345	0.333	0.323	0.315	0.307	0.301	0.295	0.289
450	0.738	0.507	0.426	0.384	0.357	0.338	0.323	0.312	0.302	0.294	0.287	0.281	0.276	0.271	0.267
500	0.665	0.457	0.386	0.348	0.325	0.308	0.296	0.286	0.278	0.271	0.266	0.261	0.257	0.253	0.250
550	0.605	0.417	0.353	0.319	0.298	0.284	0.273	0.265	0.258	0.253	0.249	0.245	0.242	0.240	0.237
600	0.555	0.384	0.325	0.295	0.277	0.264	0.255	0.248	0.243	0.239	0.235	0.233	0.231	0.229	0.228
650	0.513	0.355	0.302	0.275	0.259	0.248	0.240	0.235	0.230	0.227	0.225	0.223	0.222	0.221	0.220
700	0.477	0.331	0.283	0.258	0.244	0.234	0.228	0.223	0.220	0.218	0.216	0.215	0.214	0.214	0.214
750	0.446	0.311	0.266	0.244	0.231	0.223	0.217	0.214	0.211	0.210	0.209	0.209	0.209	0.209	0.210
800	0.419	0.293	0.251	0.231	0.220	0.213	0.208	0.206	0.204	0.203	0.203	0.204	0.205	0.206	0.207
850	0.395	0.277	0.238	0.220	0.210	0.204	0.201	0.199	0.198	0.198	0.199	0.200	0.201	0.203	0.205
900	0.374	0.263	0.227	0.211	0.202	0.197	0.194	0.193	0.193	0.194	0.195	0.197	0.199	0.201	0.203
950	0.355	0.250	0.217	0.202	0.195	0.191	0.189	0.189	0.189	0.190	0.192	0.195	0.197	0.199	0.201
1000	0.338	0.239	0.208	0.195	0.188	0.185	0.184	0.185	0.186	0.188	0.190	0.193	0.195	0.197	0.198
1100	0.308	0.220	0.193	0.183	0.178	0.176	0.177	0.178	0.181	0.184	0.187	0.190	0.192	0.193	0.194
1200	0.284	0.204	0.181	0.173	0.170	0.170	0.171	0.174	0.178	0.182	0.185	0.187	0.189	0.190	0.191

Table 1 — All Wheel-Type Tractors

Table 1 contains accumulated average fixed and repair costs for all wheel-type tractors. Look at the cost figure shown opposite 350 hours of annual use under the five-years-of age column. This figure is $0.451 (Table 1). At the top of the table, note that cost figures are for each $1,000 of list price. The costs included in these tables are:

- Depreciation
- Taxes
- Shelter
- Interest
- Insurance
- Repairs

Because the tractor had a new list price of $130,000, it would have 130 cost units of $1,000 each.

$130,000 (new list price) divided by $1,000 = 130 cost units

Therefore, multiply the cost figure from the table by 130:

130 costs units x $0.451 = $58.63 per hour of use

With 5 years of use at 350 hours per year, the accumulated use of the $130,000 tractor would be 1,750 hours. Total expenditures for fixed cost and repairs would be:

1,750 hours x $58.63 per hour = $102,602

Estimating Machine Cost

All of the cost tables in the Appendix can be used in the same way. Fuel and lubricant costs must also be computed for tractors and implements that are engine driven (Fig. 3). Review the methods for estimating fuel and lubricant costs contained in Chapter 7.

Fig. 3 — For Machines With Engines, Fuel and Lubricant Costs Must Also Be Included in Cost Estimates

The average fuel consumption for diesel engines is 0.044 gal/hp-hr times the rated PTO horsepower. A 150 horsepower diesel engine tractor would use 6.6 gallons per hour (150 hp x 0.044 gal/hp-hr) as an average for the whole year. The average fuel consumption for diesel engines, using metric values, is 0.223 L/kWh times the rated PTO kilowatt (see Appendix, Table 17). A 111.9 kilowatt diesel-engine tractor would use 25.0 liters per hour (111.9 kW x 0.223 L/kWh) as an average for the whole year.

Now calculate fuel and lubricant costs:

Fuel cost = 6.6 gallons per hour x $4.00 per gallon
= $26.40 per hour

Fuel cost (M) = 25.0 liters per hour x $1.06 per liter
= $26.50 per hour

Lubricant cost = 10% of fuel cost
= 0.10 x $26.40 = $2.64 per hour
(M) = 0.10 x $26.50 = $2.65 per hour

Total fuel and lubricant costs = $26.40 + $2.64 = $29.04 per hour
(M) = $26.50 + $2.65 = $29.15 per hour

For 1,750 hours of use, the total fuel and lubricant cost would be $46,200 (1,750 x $26.40).

Including fixed costs, repair costs, and operating costs, the average cost for the $130,000, 150-horsepower (111.9-kilowatt) tractor over a five-year period would be:

Fixed cost plus repairs = $58.63 per hour

Fuel and lubricant costs = $29.04 per hour

Total = $87.67 per hour

Not including labor, average annual cost is:

350 hours per year x $87.67 per hour = $30,684.50 per year

Fixed Costs Plus Repairs = $58.67 per Hour
Fuel and Lubricant Costs = $29.04 per Hour
Average Cost = $87.67 per Hour

Fig. 4 — Total Costs for a $130,000 Tractor Used 1,750 Hours in 5 Years

Thus, the average total cost for the tractor after 5 years of use at 350 hours per year would be $87.67 per hour, not including labor (Fig. 4).

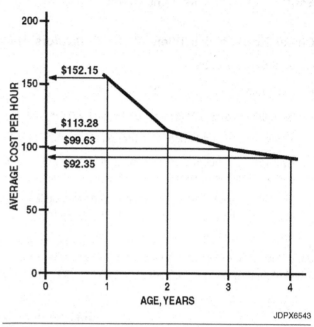

Fig. 5 — Average Cost per Hour for a $130,000 Tractor Used 350 Hours per Year

Try estimating the average cost at the end of each of the first 4 years for the same tractor. The answers should be (Fig. 5):

Year 1 = $152.15 per hour

Year 2 = $113.28 per hour

Year 3 = $99.63 per hour

Year 4 = $92.35 per hour

In using the tables in the Appendix for estimating cost, remember that only costs for repairs, depreciation, taxes, shelter, insurance, and interest are included. For total cost, add fuel and lubricant costs plus labor if it is to be included also.

Let's try another example of calculating total cost.

What is the average cost for a $12,000 mulch tiller after 10 years with 100 hours per year of use? Capacity is 7.0 acres (2.83 hectares) per hour.

Shown in Table 2 and Appendix, Table 7 is a cost of $1.463 per hour per $1,000 of list price for the mulch tiller. Round this number off to $1.46.

Dividing the list price of $12,000 by 1,000 gives a factor of 12.0, the number of $1,000 units.

Multiplying the 12.0 units by $1.46 per hour provides an average cost of $17.52 per hour for the mulch tiller. The mulch tiller by itself would not have any fuel or lubricant cost.

Estimating Tractor-Machine Costs

Chisel Plows, Mulch Tillers, Disks, Cultivators, Harrows, etc.

- Useful life = 2,000 hours
- Repair costs = $1,000 \times 0.26524 \times (hours/1,000)^2$
- Remaining value. RV ($) = $1,000 \times 0.67 \times 0.92^y$
- Taxes, shelter, insurance, and interest ($) = $0.13 \times RV$
- Average accumulated cost per hour per $1,000 of list price (includes fixed costs and repairs, but does not include labor or fuel and lubricant costs)
- Machine is assumed to be purchased new at 85% of list price. Figures shown are average accumulated costs based on original ownership since machine was purchased.

The underlined portion of the table indicates that total machine usage exceeds the useful life. When useful life is reached, the annual repair cost is carried forward at the same rate for each of the following years. Values in this table do not include the effect of inflation over the period of ownership. Use Table 7, Appendix, to estimate costs for mulch tillers.

Cost per Hour per $1,000 of List Price

Hours of Annual Use	Age, Years (y)														
	1	2	3	4	5	6	7	8	9	10	11	12	13	14	15
20	17.260	11.912	9.963	8.874	8.135	7.578	7.128	6.749	6.421	6.130	5.869	5.631	5.413	5.212	5.026
40	8.676	6.016	5.052	4.516	4.154	3.882	3.663	3.479	3.320	3.179	3.053	2.938	2.834	2.737	2.647
60	5.821	4.060	3.426	3.076	2.841	2.664	2.523	2.405	2.303	2.213	2.133	2.060	1.993	1.932	1.875
80	4.398	3.087	2.619	2.363	2.191	2.064	1.962	1.877	1.804	1.740	1.683	1.631	1.584	1.541	1.501
100	3.547	2.507	2.139	1.939	1.807	1.709	1.631	1.567	1.512	1.463	1.420	1.382	1.346	1.314	1.284
120	2.981	2.123	1.822	1.661	1.555	1.477	1.415	1.365	1.321	1.284	1.250	1.220	1.193	1.169	1.146
140	2.579	1.851	1.599	1.464	1.377	1.314	1.264	1.224	1.189	1.159	1.133	1.109	1.088	1.069	1.051
160	2.278	1.648	1.432	1.319	1.246	1.194	1.153	1.121	1.093	1.069	1.048	1.029	1.013	0.997	0.981
180	2.045	1.492	1.305	1.208	1.147	1.103	1.070	1.043	1.020	1.001	0.985	0.970	0.956	0.942	0.928
200	1.860	1.368	1.204	1.120	1.068	1.032	1.004	0.982	0.964	0.949	0.934	0.920	0.906	0.892	0.880
220	1.709	1.267	1.123	1.050	1.006	0.975	0.952	0.935	0.920	0.908	0.896	0.884	0.873	0.862	0.851
240	1.584	1.184	1.056	0.992	0.954	0.929	0.910	0.896	0.885	0.873	0.862	0.851	0.840	0.829	0.819
260	1.478	1.115	1.000	0.945	0.912	0.891	0.876	0.865	0.854	0.842	0.831	0.820	0.810	0.800	0.790
280	1.388	1.056	0.953	0.904	0.877	0.860	0.848	0.839	0.830	0.821	0.811	0.802	0.793	0.784	0.776
300	1.311	1.005	0.913	0.870	0.847	0.833	0.824	0.814	0.804	0.795	0.785	0.776	0.767	0.758	0.749
320	1.243	0.962	0.878	0.841	0.822	0.811	0.804	0.797	0.788	0.780	0.772	0.764	0.756	0.748	0.740
340	1.184	0.924	0.848	0.816	0.800	0.792	0.783	0.773	0.764	0.755	0.746	0.737	0.729	0.721	0.713
360	1.132	0.890	0.822	0.795	0.782	0.776	0.768	0.761	0.753	0.745	0.737	0.729	0.721	0.714	0.707
380	1.086	0.861	0.799	0.776	0.766	0.762	0.756	0.750	0.743	0.736	0.729	0.722	0.715	0.709	0.703
400	1.044	0.835	0.779	0.759	0.751	0.743	0.735	0.726	0.717	0.709	0.701	0.694	0.686	0.679	0.673

Table 2 — Chisel Plows, Mulch Tillers, Disks, Cultivators, Harrows, etc.

Since costs for the $130,000 tractor and the $12,000 mulch tiller have been calculated, we can combine the two for estimating the total cost of an operation such as mulch tillage.

Tractor cost = $87.67 per hour

Mulch tiller cost = $17.52 per hour

Total cost for operation = $105.19 per hour

Fig. 6 — Total Cost for the Mulch Tiller Operation in This Situation Is $105.19 per Hour or $15.03 per Acre ($33.33 per Hectare)

If the average capacity for the mulch tilling operation is 7.0 acres per hour, then dividing $105.19 by 7.0 acres per hour provides a cost per acre of $15.03 (Fig. 6). In metric equivalents, average capacity is 2.83 hectares per hour, which equates to a cost of $37.17 per hectare. But using these cost tables involves two important assumptions.

One assumption is the average annual use. A second assumption is the age of the machine or how long it is planned to own the machine. In the previous example, we assumed the tractor is owned 5 years with a use of 350 hours per year.

How would the average cost per hour be changed for the tractor if it is used 600 hours per year for 5 years? Go back to Table 1 or the tables in the Appendix and calculate the tractor cost for 600 hours per year for 5 years.

Information in the table shows fixed cost plus repairs to be $0.277 per hour per $1,000 list price.

$130,000 divided by $1,000 = 130 cost units

130 cost units x $0.277 = $36.01 per hour for fixed cost plus repairs

Adding the fuel and lubricant costs of $29.04 per hour to fixed costs plus repairs, we can get the total cost by adding as follows:

Fixed cost plus repairs = $36.01 per hour

Fuel and lubricant costs = $29.04 per hour

Total cost = $65.05 per hour

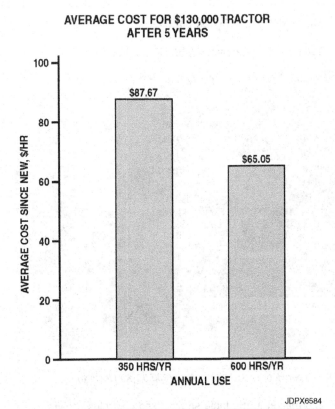

Fig. 7 — Increasing Annual Use Reduces Cost per Hour

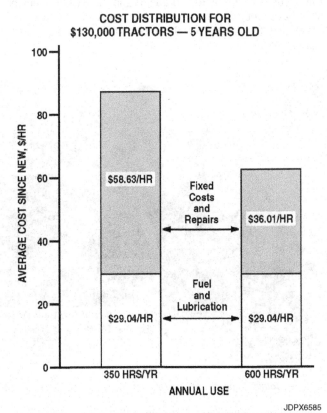

Fig. 8 — How Various Costs Are Affected by Amount of Use

Increasing tractor use to 600 hours per year in this example cuts the average cost from $87.67 to $65.05 per hour (Fig. 7). When added to the tiller cost, assuming annual use of the mulch tiller remains at 100 hours per year, we arrive at the total cost:

Tractor cost = $65.05 per hour

Mulch tiller cost = $17.52 per hour

Total cost for operation = $82.57 per hour

At 7.0 acres per hour, mulch tilling cost would be $11.80 per acre ($82.57/hr divided by 7.00 acres/hr), not including labor.

At 2.8 hectares per hour, mulch tilling costs would be $29.49 per hectare ($82.57/hr divided by 2.8 ha/h), not including labor.

These examples show that the more a machine is used, the lower the cost per unit of use. Increasing tractor use from 350 to 600 hours per year lowered cost from $105.19 to $82.57 per hour for a savings of $22.62 per hour.

Fuel and lubricant costs tend to remain constant on a hourly basis, while repair costs are almost directly proportional to the amount of annual use (Fig. 8). Fixed costs, however, are indirectly proportional to use.

Estimating Costs for Combines

One of the most widely used machines is a self-propelled combine, so we will use this machine to show another important principle — estimating combine costs.

Suppose we want to determine the average cost per acre for a $225,000 corn combine that has a capacity of 6 acres (2.43 hectares) per hour. This time we will include labor cost of $10.00 per hour. Diesel fuel cost is $4.00 per gallon ($1.06 per liter).

Assuming an annual use of 240 hours, what is the average cost per acre at the end of 3, 6, 9, and 12 years?

Data in Table 3, in the Appendix, show fixed cost plus repairs. Using this information the costs for the combine can be determined as shown in Table 3.

Cost Item	Age, Years			
	3	6	9	12
Fixed and Repair Cost, $ Per $1,000 of List Price (From Table 3, Appendix)	0.809	0.658	0.606	0.584
Fixed and Repair Cost, $225,000 Combine, $/Hour	182.02	148.05	136.50	131.40
Labor Cost, $/Hour	10.00	10.00	10.00	10.00
Fixed, Repair and Labor Cost, $/Hour	192.03	158.05	146.50	141.40
Fuel and Lubricant Cost, $/Hour	42.24	42.24	42.24	42.24
Total Cost, $/Hour	234.27	200.29	188.74	183.64
Total Cost, $/Acre	25.80	22.32	21.13	20.61

Table 3 — Corn Combine Costs

Fig. 9 — The Effect of Age on Average Costs for a $140,000 Corn Combine

The effect of length of ownership is illustrated in this example (Fig. 9).

System Costs

We have shown how to estimate total cost for a single machine or tractor-machine combination. Next, we will look at the complete system in Table 4 to show the division of costs and total cost for an entire operation. The last section of this manual will provide a better understanding of calculating total cost for a complete operation.

Fig. 10 — Costs for Entire System Can Be Estimated With Information in This Manual

Operation	Field Capacity	New Cost	Hours per Year	Years to Own
Self-Propelled Swather	6.6 acre/hr 2.67 ha/hr	$50,000	160	10
Twin Rakes	8 acre/hr 3.24 ha/hr	$12,500	140	10
Large Roll Baler	8 acre/hr 3.24 ha/hr	$23,000	140	10
Hauling	($5.00/ton charge assumed)			

Table 4 — Costs for Hay Operation

The example shown in Table 4 is for a hay operation, using a large roll baler, with each bale weighing 1 ton (Fig. 10).

Assume the yield to be 1.5 tons per acre, labor is $10.00 per hour, baling requires a tractor at $45.60 per hour, and swathing requires 0.55 gallons of diesel fuel per acre. Diesel fuel is $4.00 per gallon. Twine for the bales costs $1.50 per bale.

Raking requires a tractor at $45.60 per hour. Hauling the large roll bales to storage is assumed to cost $5.00 per bale.

Operation	$/Hour	$/Acre	$/Hectare	$/Ton
Swathing	68.24	10.34	25.56	6.89
Raking	68.68	8.59	21.46	5.73
Bailing	106.35	13.29	32.82	8.86
Hauling		7.50	18.53	5.00
		39.72	92.82	26.48

Table 5 — Cost Summary

Ⓜ Assume the yield is 3.36 metric tons per hectare, each bale weighs 0.91 metric tons, labor is $10.00 per hour, baling requires a tractor at $45.60 per hour, swathing requires 5.1 liters of fuel per hectare, and diesel fuel is $1.06 per liter. Twine costs $1.50 per bale.

Summary

This chapter addresses estimating total cost for machines, operations, and systems. Accurate records are the best source of material for estimating costs, but cost tables are provided in the Appendix for estimating machine costs. The costs included in these tables cover:

- Depreciation
- Taxes
- Shelter
- Interest
- Insurance
- Repairs

Fuel and lubricant costs must be included when estimating total cost for machines. Methods for estimating fuel and lubricant costs were provided in Chapter 7.

Labor costs should also be included. Labor can be easily added for each situation. With fuel, lubricant, and labor costs added to the fixed and repair costs included in the tables, the total cost for an entire system can be calculated. Estimating total cost for a system is a basic skill for machinery management and an integral part of the following chapters.

Why It Is Important to Know Total Cost

Every decision relating to machinery should be based on accurate total cost estimates. Knowing the average cost per hour or acre illustrates either an increase or decrease. This information can be used for decisions regarding trading or purchasing equipment. When repair costs become too large, it might be time for new equipment. Knowledge of total cost is of greatest importance when comparing ownership of a machine to hiring a custom operator.

Test Yourself

Questions

1. List three important machinery management decisions you could make if you know how to make an accurate cost estimate.

2. What costs are included in the "cost per hour" tables?

3. What are the average annual fixed and repair costs per hour of use for each $1,000 of new cost for a tractor that is used 800 hours per year for 4 years of ownership? (Use tables in Appendix.)

4. What is the average annual total cost per hour for a $42,200, 80-horsepower tractor for 4 years if it is used 800 hours per year? (Assume diesel fuel at $4.00 per gallon.)

5. If labor is $10.00 per hour, what is the average annual cost per hour for a $45,000 self-propelled swather for 4 years of ownership if it is used on 500 acres per year, and operates at 5.0 acres per hour? It has a diesel engine that consumes 5.2 gallons of fuel per hour. Diesel fuel cost is $4.00 per gallon. Lubricant cost equals 10% of fuel cost.

6. What is the average annual cost per acre for a self-propelled combine with a list price of $225,000 if it is used for 700 acres of soybeans and 700 acres of corn a year? Labor is $10.00 per hour. Diesel fuel is $4.00 per gallon. Lubricant costs equal 10% of fuel cost. Capacity for the combine is 6.5 acres per hour in soybeans and 6 acres per hour in corn. Assume that the combine will be owned 8 years. (Hint: You will need to use some of the tables and information in previous chapters to solve this problem.)

7. What are three ways to lower the cost per acre for machinery?

8. Select a system involving three or more operations and determine the total system cost. Make your own assumptions for years to own, annual use, labor cost, etc.

III
Managing Machinery

	Chapter
Deciding When to Trade	10
Considering Future Capacity Needs	11
Calculating Custom Work Costs	12
Decision Time — Selecting the Best Alternative	13
Case Studies in Machinery Management	14

Managing Machinery

In the first five chapters (Section I), basic information was presented about machine performance, speed, field efficiency, and capacities. Information was also provided on how to match power units and machines for satisfactory field operation.

The four chapters in Section II showed how some simple but meaningful cost estimates can be made for individual machines or combinations of machines.

In Section III, this background information will be used to show how to make those important, moneymaking decisions. This will not be a simple matter of looking up the right answer for every tough question that comes along. Many solutions have to be thought through and calculated.

Important machinery management decisions are not easy. In fact, most authorities agree that machinery management decisions are harder to pin down than other decisions related to farming.

Most machinery management problems can be solved by doing the following:

- Realistically estimate total working time available for major field operations
- Determine the required effective field capacity of machines
- Match power units to machines
- Predict costs for machine application

These final chapters demonstrate combining the basic ideas for making management decisions. Reference data presented are intended to be as accurate as any available. However, it may not always be perfect for each situation. One should adapt this reference information to solve specific problems.

Carefully follow through with the sample problems in these final chapters. Doing so will greatly improve machinery management ability or provide a good review for those more experienced as machinery managers.

Deciding When to Trade

10

Introduction

In the early 1990s, cost of farm machinery steadily increased without a corresponding increase in prices for farm products. Longer machinery ownership, with increased emphasis on maintenance and repairs, helps to compensate for the problem of the cost-price squeeze.

Fortunately, farm machinery is built to last. Records show combines lasting well beyond three thousand hours and tractors as much as twelve thousand hours. Except for extremely high annual use, most machines do not reach the point of minimum cost before the 15th year. However, if machinery is kept too long, repair costs increase and reliability is lost (Fig. 1).

Chapter Objectives

- Establish trading guidelines

Establishing Trading Guidelines

Good guidelines are important for making management decisions on when to trade machinery (Fig. 1). Five important reasons for trading a machine are:

- The accumulated average annual cost per unit of use has reached its lowest point and is increasing.

- The machine is obsolete in comparison to new models.

- The machine has lost its reliability, meaning it is no longer dependable.

- The machine is worn out.

- The increase in the size of an operation makes a machine too small for timely operations.

Fig. 1 — Determining When to Trade Should Be Based on Facts for the Best Profits

In all cases, investment credit or other tax credits need to be included when considering the purchase of new machinery. It is impossible to address these issues in this book due to the constant changes in tax laws. Instead it is suggested that you first learn how to make an accurate cost analysis, and then determine how it is affected by tax laws. When you do this, you will have a strong foundation on which you can make decisions.

Average Cost per Unit of Use

Accumulated costs are those that include all of the costs that are incurred since the date of ownership. On the other hand, annual costs refer only to the costs for one year. All of the annual costs added together would give the accumulated costs. Dividing the accumulated costs by the years of ownership gives the average annual costs.

Typical average accumulated costs for tractors and machinery are shown in Fig. 2. Notice how the average cost per hour (since the time of purchase) begins high, decreases rapidly for a few years, and then gradually decreases. With typical operating conditions, the point of minimum cost will seldom occur before the 10th year.

In Chapter 6, we saw that fixed costs decrease with age. When the average annual repair costs start increasing faster than average annual fixed costs are decreasing, the accumulated average cost will start rising.

The annual cost curve as well as the average accumulated cost curve for a $90,000 tractor used 800 hours a year are shown in Fig. 3. The curves intersect in our example at about the 11th year.

Theoretically, the time to trade is when the annual cost starts to exceed the average accumulated cost over the entire time the machine has been owned. This rule is right in cases where the annual use is high and extensive repairs make the annual cost quite high.

But, it is not always advisable to trade when annual cost equals the average accumulated cost. For example, tractors having an annual use of less than 800 hours will not reach their point of minimum cost until well beyond the 10th year.

Normally, planning to own any machinery or tractor more than 10 years can be risky due to the danger of obsolescence, high repair bills, and loss of reliability. However, when you are financially squeezed or simply cannot afford to trade, then the best alternative is to continue to own. In this case, you must take care to properly maintain the machine (Fig. 4).

Also, average accumulated costs tend to flatten after about 6 years (Fig. 2). The difference in the average cost per hour for a $90,000 tractor used 800 hours a year is not very great after the tractor is 6 years old. The cost after 6 years would be $0.85 per hour higher, compared to 10 years (Fig. 5).

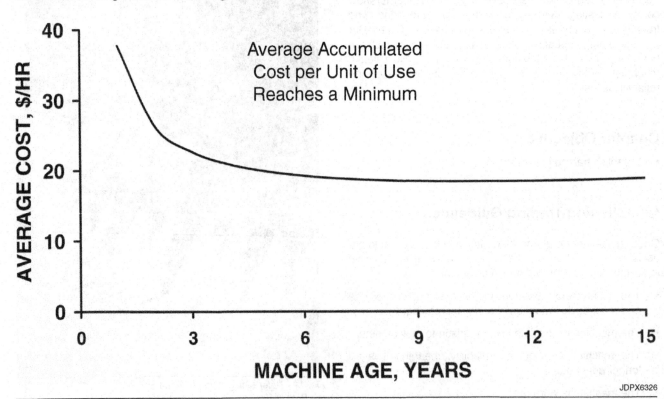

Fig. 2 — One Basis for Trading Is When the Average Cost per Unit of Use Has Reached Minimum

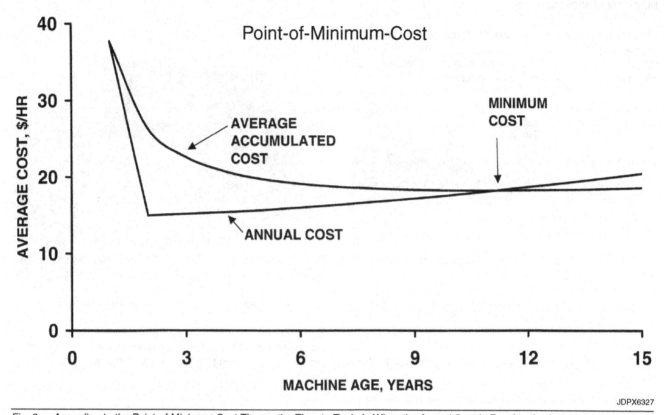

Fig. 3 — According to the Point-of-Minimum-Cost Theory, the Time to Trade Is When the Annual Cost Is Equal to the Accumulated Average Cost per Unit of Use

Fig. 4 — Older Machines Need Better Care, Repair, and Maintenance

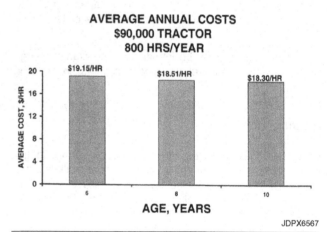

Fig. 5 — Cost Savings From Long Ownership May Not Be Justified Due to Increased Odds of Loss of Reliability

Machine Obsolescence

Machines, and to some extent tractors, can become obsolete before they are worn out completely (Fig. 6). Obsolete means a machine is no longer as well suited for the work to be accomplished as newer models.

Fig. 6 — Machines Can Become Obsolete Before They Are Worn Out

Tractors are basically power units and therefore do not become obsolete as fast as combines, balers, swathers, and other machines.

Machines can be considered to become obsolete in three ways:

- A new model of the same machine may have a design change resulting in better efficiency and increased capacity.
- Increased power and the resulting capacity may be needed due to expanded field operations or less labor available.
- A new concept may be introduced. For example, large square balers are relatively new machine concepts in hay handling.

Machine Reliability

When a machine loses its reliability, it is no longer dependable for completing the job in the time available. It is difficult to repair a machine and economically regain an acceptable level of reliability once periodic breakdowns occur.

If a machine starts breaking down often, the amount of repairs needed to restore an acceptable level of reliability may be prohibitive. This is especially true for the larger, more complex machines.

It is difficult to determine exactly when a machine is no longer reliable. The greater the amount of annual use or the larger the capacity of the machine, the higher the reliability must be (Fig. 7).

Fig. 7 — The Greater the Use, the More Important Is Reliability

By maintaining a high level of reliability through good maintenance and correct operation, you can count on a machine's use during 95% or more of the available working time.

Worn-Out Machinery

Finally, trade a machine when it is worn out or when it simply will not perform its job properly even with repairs. Here again, there is a tremendous difference in the total time from new to worn out. It depends on the kind of care the machine has had and how it is used. Proper daily care, periodic servicing, and using care not to overload or abuse a machine can double or triple its life (Fig. 8).

Fig. 8 — Proper Care and Maintenance Can Double or Triple Machine Life

Calculating Machine Life

Examine typical or average situations to establish useful guidelines for wear-out, least-cost, or obsolescence life. Data in Table 1 are available for that purpose.

Machine	Wear-Out Life, Hours	Least-Cost Life, Hours
All Wheel Tractors	12,000	8,000
Crawlers	16,000	10,000 to 13,000
Self-Propelled Combines	3,000	3,000
Cotton Harvesters	5,000	5,000
Large Square Balers	3,500	2,500
Planters, Drills	1,500	1,100
Moldboard Plows	2,000	1,400
Disk, Chisels, Harrows, Cults, etc.	2,000	2,000
Mowers	2,000	1,100
Small and Large Square Balers	2,500	2,500
Large Round Balers	1,500	1,200
Self-Propelled Forage Harvesters	4,000	4,000
Self-Propelled Windrowers	3,000	3,000
Rakes	2,500	2,500

Table 1 — Average Machine Wear-Out and Least-Cost Life

The information in Table 1 is only a guideline that indicates approximate wear-out and least-cost life. It is intended to show that extending ownership will continue to lower the cost per unit of use, as long as an acceptable level of reliability can be maintained.

Some machines have relatively high repair costs and/or short least-cost life compared to wear-out life. Included in this category are tractors, planters, drills, mowers, and balers. For those high annual use situations requiring equipment reliability, it might be feasible to trade at an earlier time than indicated in Table 1. One-half to three-fourths of wear-out life should be used.

For example, 200 hours a year would be considered high annual usage for a corn planter. Table 1 shows a wear-out life of 1,500 hours. Three-fourths of 1,500 hours would be 1,125 hours, or a little over 5 years of use.

Again, Table 1 is only a suggested guideline. If the planter still has a high level of reliability and is not obsolete, then continued ownership is the best alternative, since cost per acre will continue to decrease.

Obsolescence life of any machine is difficult to predict, since new designs may come along at any time. A tractor or machine sized properly for the amount of work to be done should have high annual usage and will reach its wear-out or least-cost life before becoming obsolete.

Concepts of new machines are difficult to judge and may require trial and error. It is better to plan to trade most new concept machines in 6 years or less. This reduces the risk of owning a machine for an extended period that proves unfeasible for your situation.

When to Repair

Whether to trade or repair is a problem for machinery owners when they are limited on capital or want to cut costs without hurting production. You can expect repair costs to increase with annual use and age (Table 2 and Table 3).

	Annual Hours of Use					
Year	200	300	400	500	600	700
	Repair Costs per Year per $1,000 of List Price					
1	$0.28	$0.62	$1.11	$1.74	$2.50	$3.40
2	$0.83	$1.87	$3.33	$5.21	$7.50	$10.21
3	$1.39	$3.12	$5.56	$8.68	$12.50	$17.01
4	$1.94	$4.37	$7.78	$12.15	$17.50	$23.82
5	$2.50	$5.62	$10.00	$15.62	$22.50	$30.62
6	$3.06	$6.87	$12.22	$19.10	$27.50	$37.43
7	$3.61	$8.12	$14.44	$22.57	$32.50	$44.23
8	$4.17	$9.37	$16.67	$26.04	$37.50	$51.04
9	$4.72	$10.62	$18.89	$29.51	$42.50	$57.84
10	$5.28	$11.87	$21.11	$32.98	$47.50	$64.65
11	$5.83	$13.12	$23.33	$36.46	$52.50	$71.45
12	$6.39	$14.37	$25.55	$39.93	$57.50	$78.26
13	$6.94	$15.62	$27.78	$43.40	$62.50	$85.06
14	$7.50	$16.87	$30.00	$46.87	$67.50	$91.87
15	$8.06	$18.12	$32.22	$50.34	$72.50	$98.67

Table 2 — Annual Repair Costs for Tractors

	Annual Hours of Use					
Year	50	100	150	200	250	300
	Repair Costs per Year per $1,000 of List Price					
1	$0.07	$0.32	$0.74	$1.36	$2.17	$3.18
2	$0.24	$1.04	$2.44	$4.46	$7.12	$10.44
3	$0.42	$1.82	$4.27	$7.81	$12.48	$18.29
4	$0.61	$2.64	$6.18	$11.30	$18.06	$26.48
5	$0.81	$3.47	$8.14	$14.90	$23.80	$34.91
6	$1.01	$4.33	$10.15	$18.58	$29.68	$43.53
7	$1.21	$5.21	$12.20	$22.32	$35.67	$52.30
8	$1.42	$6.09	$14.28	$26.13	$41.74	$61.22
9	$1.63	$6.99	$16.39	$29.98	$47.91	$70.25
10	$1.84	$7.90	$18.52	$33.88	$54.14	$79.40
11	$2.06	$8.82	$20.67	$37.83	$60.44	$79.40
12	$2.27	$9.75	$22.85	$41.81	$66.80	$79.40
13	$2.49	$10.69	$22.05	$45.82	$66.80	$79.40
14	$2.71	$11.63	$27.26	$49.87	$66.80	$79.40
15	$2.94	$12.58	$29.49	$53.95	$66.80	$79.40

Table 3 — Annual Repair Costs for Combines

Tractors and other machines seldom reach their point of minimum cost before the 10th year. There may be many reasons, including loss of reliability, that justify the trade before that time. If you are in a tight squeeze and cannot afford the trade, these are the factors that should be considered. A good maintenance program will reduce expenses. Avoid overloading and lugging down engines. Properly match machines and equipment (Chapter 5). These practices will help you avoid expensive repair costs to the engine and power train.

You can save expenses by rigidly following a good maintenance program.

When to Repair and Continue Using

Avoid trading in machinery in the following cases:

- When the machine or tractor has a good record of reliability.
- When you have followed a good maintenance program, and repairs have been only those expected (Chapter 8).
- When there have been no problems with the engine and power train other than normal overhauls.
- When equipment is adequate for your farming needs.
- When the equipment is fairly simple, such as tillage tools, it is more economical to repair as long as the machine is functional and does not have a damaged frame or bent axle.

When to Trade

Consider trading in the following situations:

- When the tractor or machine has had a major failure in the power train, such as the transmission, final drive, or ring gear.
- When the equipment breaks down frequently.
- When extensive repairs are required on a machine that is not properly suited to your needs.
- When there is an opportunity to purchase or trade for a good used machine.

These guidelines may work for some situations. Consider situations where a more complete cost analysis is needed.

Example: You have a 6-year-old diesel tractor with 3,600 hours on it. You know it will need extensive injection system repairs, new tires, and some minor repairs. It cost $70,000 new. You can trade for a new one for $30,000. What should you do?

Step 1. Assume the comparison for the next 4 years, which would take the tractor to 10 years of age.

Step 2. Estimate the cost of repairs for your tractor for the next 4 years (Table 2).

The estimated repair costs for the $70,000 tractor for the next 4 years would be:

$70,000 divided by 1,000 = 70 cost units for Table 2.

70 x (32.50 + 37.50 + 42.50 + 47.50) = $11,200

The $11,200 figure represents an estimate for repairs for the old tractor for the next 4 years, and should cover all of the anticipated repairs listed earlier. In terms of cash flow and financial considerations, it will cost about one third as much to repair than to trade. The new tractor would, however, start another 10 years or more of service life.

Consider another example:

Suppose that the same tractor needs a transmission overhaul that costs $6,000. Add this to the $11,200 of anticipated repairs over the next 4 years. This gives a total of $17,200, which is still well under the cost to trade.

However, the transmission failure may be a signal that other major failures of the power train might occur in the next 4 years. Each of these would be very expensive to make on an old tractor with little resale value. In this case, you may save expenses if you sell the tractor and look for a reliable used one.

Table 3 gives the average annual repair costs for combines. Table 1 in Chapter 8 can be used to make estimates for other machines.

Suppose that you have a $140,000 list price combine that has been used 8 years at 250 hours a year. What should you expect to pay for repairs over the next 2 years?

Look at Table 3. The anticipated repair costs for the 9th year would be $47.91 per $1,000 of list price and $54.14 for the 10th year.

140 cost units x ($47.91 + $54.14) = $14,287

The cost to trade would be considerable and you would have to review the reliability of the combine after it is 8 years old. Many combines have low annual hours of use. They must, however, be reliable and not be out of service during critical harvest periods.

Summary

The most important factor in trading machinery for the best profits is to base trades on facts. The four important reasons for trading machinery discussed in this chapter are:

- The average annual cost per unit of use is close to or has reached its minimum point.
- The machine is obsolete.
- The machine is unreliable.
- The machine is worn out.

An increase in the size of an enterprise or a desire to take advantage of the first-year depreciation in investment tax credit are two other reasons to trade machinery. These are discussed in previous chapters.

For most machines, the earliest feasible trade time occurs at approximately three-fourths of the least-cost life. Certain machines may have an earlier feasible trade time, depending on the rate of usage. When buying a machine that is a new concept, use a shorter obsolescence life of about 6 years.

Test Yourself

Questions

1. List four main reasons why a machine should be traded.
2. How do you determine when a machine should be traded on the least-cost basis?
3. When is a machine no longer reliable?
4. What is the suggested obsolescence life for wheel tractors? Combines? Planters?
5. A tractor is used 1,000 hours per year. When should it be traded? Explain why.
6. What are the reasons why you should repair a used machine rather than trade for a new one?
7. How much would the repair costs be for a tractor with a $90,000 list price for years 6 through 8 of ownership if it is used 500 hours a year?
8. What is the biggest problem in owning a combine that has a history of major repairs?

Considering Future Capacity Needs

11

Introduction

Fig. 1 — Buy the Size That Will Get Important Jobs Done on Time

An important aspect of buying a new machine or tractor is making the decision based on the machine size needed. Buying a tractor or machine with adequate power or capacity is a good idea, even if financing the purchase is necessary (Fig. 1).

If decisions are based on available cash, a tractor purchased could end up being used only 100 hours per year, or an undersized tractor may require too much time to complete a job. Other alternatives to ownership of machinery are available and will be covered in later chapters.

Another potential basis for machine selection is to purchase equipment large enough to allow one person to do all of the work. This might be the best choice for smaller, less diversified operations.

Chances are, however, that managing larger, more diversified farming operations require more than one person to perform several ongoing jobs. The involvement of hired labor on expensive machinery usually means some type of supervision and the need for the manager to be available for on-the-spot decisions.

The balance between labor and machine size is an important one. Certain types of agricultural enterprises may require a large amount of hired labor, while labor requirements on other types may be reduced by larger machines. The important thing is that laborsaving dollars might be eliminated or surpassed by high fixed costs of large machines. Consider these factors when selecting machinery or tractor size.

An integral concept is tractor or machine size selection regarding future increases in the size of an agricultural operation.

Chapter Objectives

- Select tractor or machinery size needed
- Calculate tractor or machinery costs
- Match tractor and machine to allotted time to complete operations

Selecting Tractor Size

A logical approach to deciding how large a tractor should be is to ensure two things:

- Provide for enough power to get all important field operations completed on time (Fig. 2).

Fig. 2 — Adequate Power Is Needed to Complete Key Operations Such as Tillage on Time

- Provide for sufficient annual use so costs will be minimized (Fig. 3).

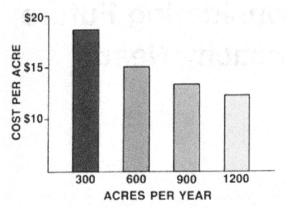

Fig. 3 — Sufficient Annual Use Is Needed to Keep Costs Down

First, list all field operations according to energy requirements and estimate the total time available. For example, information in Table 1 of Chapter 5 shows that plowing has the highest energy requirement. In the following example, assume 400 acres (162 hectares) are to be plowed. Also assume the time available for plowing is 138.8 hours, as figured in Table 1.

Calendar Periods	Average Working Days	Average Hours per Day	Total Hours Available
Oct. 15–31	8.0	8.0	64.0
Nov. 1–15	6.5	7.5	48.8
Nov. 16–30	4.0	6.5	26.0
Total Hours Available =			138.8

Table 1 — Sample Fall Plowing Schedule

If using one tractor and plow, the average plowing capacity needed would be:

$$\frac{400 \text{ acres}}{138.8 \text{ hours}} = 2.88 \text{ acres per hour}$$

$$\frac{162 \text{ hectares}}{138.8 \text{ hours}} = 1.17 \text{ hectares per hour}$$

The plow size needed in the example can be calculated using the following:

- Speed = 4.5 miles per hour (7.2 km/h)
- Field Efficiency = 80%
- Draft = 800 pounds per foot (11.68 kN/m)

Using the following formula we can determine the plowing width needed in our example:

$$\text{Width} = \frac{\text{acres per hour} \times 8.25}{\text{speed, mph} \times \text{field efficiency}}$$

$$= \frac{2.88 \text{ acres per hour} \times 8.25}{4.5 \text{ mph} \times 0.80}$$

$$= 6.6 \text{ feet (80 inches)}$$

For metric formula required to determine width, see Metric Equivalents.

A 5-bottom, 16-inch plow or equivalent width would be needed.

How much drawbar horsepower is needed to pull the same plow at 4.5 miles per hour with a draft of 800 pounds per foot? First, find the total draft of the plow:

6.6 feet x 800 pounds per foot = 5,280 pounds

Now find the drawbar horsepower required to pull the plow at 4.5 miles per hour:

$$\text{Drawbar hp} = \frac{\text{Draft, pounds} \times \text{Speed, mph}}{375}$$

$$= \frac{5280 \times 4.5}{375}$$

$$= 63.4 \text{ drawbar horsepower}$$

For metric formula used to determine drawbar kilowatts, see Metric Equivalents.

For firm soil conditions using a two-wheel drive tractor, the actual drawbar power required will be approximately 61% of the rated PTO power of the tractor (see Table 2, Chapter 5). In this case, a tractor with 104 (63.4 drawbar hp/ 0.61) rated PTO horsepower would meet the requirement.

Calculating Relative Costs

Sometimes the relative cost is an important factor. Suppose a buying decision is to be made on either a 100-horsepower tractor to be used 800 hours per year or a 125-horsepower tractor to be used 650 hours per year. Both tractors could accomplish the same amount of work.

If the list price of the 100-horsepower (75-kilowatt) tractor is $80,000 and the list price of the 125-horsepower (93-kilowatt) tractor is $100,000, which would be the most economical? Use a labor cost of $10.00 per hour and a diesel fuel cost of $4.00 per gallon ($1.06 per liter). Assume a 10-year ownership in both examples.

To better understand and compare the costs, consider the costs for each tractor separately.

Costs for 100-Horsepower (75-kW) Tractor

The 100-horsepower (75-kW), diesel-engine tractor will be used 800 hours per year for 10 years and has a list price of $80,000.

Fixed Costs Plus Repairs

Use Table 2 or the Appendix Table 1 to determine the constant for fixed costs plus repairs for a 100-horsepower (75-kW) tractor. At 800 hours per year use, the cost would be $0.203 per hour per $1,000 of list price.

The $80,000 tractor would then have 80 cost units. To calculate the average fixed and repair costs, multiply the number of cost units by the constant:

80 units x $0.203 = $16.24 per hour for fixed and repair costs

Fuel and Lubricant Costs

Estimating fuel and lubricant costs is discussed in Chapter 7. Remember that the fuel consumption multiplier for diesel-engine tractors is 0.044. The equivalent multiplier when metric figures are used is 0.223.

First, calculate the fuel cost:

100 hp x 0.044 x $4.00 per gallon = $17.60 per hour

 75 kW x 0.223 x $1.06 per liter = $17.73 per hour

Note: The metric example gives a slightly different answer due to rounding of numbers.

Next, add 10% of the fuel cost as the lubricant cost:

$17.60 + (0.10 x $17.60) = $19.36 per hour

The average cost is $19.36 per hour for fuel and lubricants for the 100-horsepower (75-kW) tractor.

Now add up the costs.

Fixed costs plus repairs	= $16.24 per hour
Fuel and lubricant costs	= $19.36 per hour
Labor	= $10.00 per hour
Total average costs	$45.60 per hour

All Wheel-Type Tractors

- Useful life = 12,000 hours

- Repair costs = $1,000 \times 0.006944 \times (hours/1,000)^2$

- Remaining value. RV (\$) = $1,000 \times 0.67 \times 0.94^y$

- Taxes, shelter, insurance, and interest (\$) = $0.13 \times RV$

- Average accumulated cost per hour per \$1,000 of list price (includes fixed costs and repairs, but does not include labor or fuel and lubricant costs)

- Machine is assumed to be purchased new at 85% of list price. Figures shown are average accumulated costs based on original ownership since machine was purchased.

The underlined portion of the table indicates that total machine usage exceeds the useful life. When useful life is reached, the annual repair cost is carried forward at the same rate for each of the following years. Values in this table do not include the effect of inflation over the period of ownership.

Hours of Annual Use	\multicolumn{15}{c}{Cost per Hour per \$1,000 of List Price — Age, Years (y)}														
	1	2	3	4	5	6	7	8	9	10	11	12	13	14	15
200	1.655	1.129	0.942	0.841	0.775	0.726	0.688	0.656	0.629	0.605	0.584	0.565	0.547	0.531	0.516
250	1.325	0.904	0.756	0.676	0.623	0.585	0.555	0.530	0.509	0.490	0.474	0.459	0.446	0.434	0.422
300	1.104	0.755	0.632	0.565	0.522	0.491	0.467	0.447	0.430	0.415	0.402	0.390	0.380	0.370	0.362
350	0.947	0.648	0.543	0.487	0.451	0.425	0.404	0.388	0.374	0.362	0.352	0.342	0.334	0.326	0.320
400	0.830	0.569	0.477	0.429	0.398	0.376	0.358	0.345	0.333	0.323	0.315	0.307	0.301	0.295	0.289
450	0.738	0.507	0.426	0.384	0.357	0.338	0.323	0.312	0.302	0.294	0.287	0.281	0.276	0.271	0.267
500	0.665	0.457	0.386	0.348	0.325	0.308	0.296	0.286	0.278	0.271	0.266	0.261	0.257	0.253	0.250
550	0.605	0.417	0.353	0.319	0.298	0.284	0.273	0.265	0.258	0.253	0.249	0.245	0.242	0.240	0.237
600	0.555	0.384	0.325	0.295	0.277	0.264	0.255	0.248	0.243	0.239	0.235	0.233	0.231	0.229	0.228
650	0.513	0.355	0.302	0.275	0.259	0.248	0.240	0.235	0.230	0.227	0.225	0.223	0.222	0.221	0.220
700	0.477	0.331	0.283	0.258	0.244	0.234	0.228	0.223	0.220	0.218	0.216	0.215	0.214	0.214	0.214
750	0.446	0.311	0.266	0.244	0.231	0.223	0.217	0.214	0.211	0.210	0.209	0.209	0.209	0.209	0.210
800	0.419	0.293	0.251	0.231	0.220	0.213	0.208	0.206	0.204	0.203	0.203	0.204	0.205	0.206	0.207
850	0.395	0.277	0.238	0.220	0.210	0.204	0.201	0.199	0.198	0.198	0.199	0.200	0.201	0.203	0.205
900	0.374	0.263	0.227	0.211	0.202	0.197	0.194	0.193	0.193	0.194	0.195	0.197	0.199	0.201	0.203
950	0.355	0.250	0.217	0.202	0.195	0.191	0.189	0.189	0.189	0.190	0.192	0.195	0.197	0.199	0.201
1000	0.338	0.239	0.208	0.195	0.188	0.185	0.184	0.185	0.186	0.188	0.190	0.193	0.195	0.197	0.198
1100	0.308	0.220	0.193	0.183	0.178	0.176	0.177	0.178	0.181	0.184	0.187	0.190	0.192	0.193	0.194
1200	0.284	0.204	0.181	0.173	0.170	0.170	0.171	0.174	0.178	0.182	0.185	0.187	0.189	0.190	0.191

Table 2 — All Wheel-Type Tractors

Costs for 125-Horsepower (93-kW) Tractor

The 125-horsepower (93-kW), $100,000 diesel-engine tractor will be used 650 hours per year over a 10-year period. Once again, let's calculate each cost involved.

Fixed Costs Plus Repairs

The constant for fixed costs plus repairs for tractors is $0.227 per hour per $1,000 of list price (Table 2). Because the tractor costs $100,000 when new, we have 100 cost units. Now, calculate the fixed and repair costs for this tractor:

100 units x $0.227 = $22.70 per hour

Fuel and Lubricant Costs

Fuel and lubricant costs are calculated as in the previous example only this time for a 125-horsepower (93-kW), diesel-engine tractor:

Fuel cost would be:

125 hp x 0.044 x $4.00 per gallon = $22.00 per hour

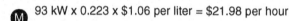 93 kW x 0.223 x $1.06 per liter = $21.98 per hour

Then, add 10% of the fuel cost as the lubricant cost:

$22.00 + (0.10 x $22.00) = $24.20 per hour

The average cost is $24.20 per hour for fuel and lubricants for 125-horsepower (93-kW) tractor.

Now add up the costs.

Fixed costs plus repairs	= $22.70 per hour
Fuel and lubricant costs	= $24.20 per hour
Labor	= $10.00 per hour
Total average costs	$56.90 per hour

Total Average Annual Costs

Next, use the total cost per hour to estimate the total annual cost for each tractor.

The 100-horsepower (75-kW) tractor is used 800 hours per year and has a total cost per hour of $45.60 per hour.

800 hours x $45.60 per hour = $36,480 per year

The 125-horsepower (93-kW) tractor is used 650 hours per year and has a total cost of $56.90 per hour.

650 hours x $56.90 per hour = $36,985 per year

By comparison, the two tractors have almost the same annual costs, including labor ($36,480 per year for the 100 horsepower tractor and $36,985 for the 125-horsepower tractor).

Without labor included, the cost would be:

100 hp (75 kW) tractor = $35.60 per hour x 800 hours per year
= $28,480 per year

125 hp (93 kW) tractor = $46.90 per hour x 650 hours per year
= $30,485 per year

Without labor, the annual cost of using the 125-horsepower (93-kW) tractor is $2,005 more per year compared to the 100-horsepower (75-kW) tractor.

This is a good example of the advantage of properly sizing the tractor for minimum costs, when each size of tractor is to accomplish the same amount of work.

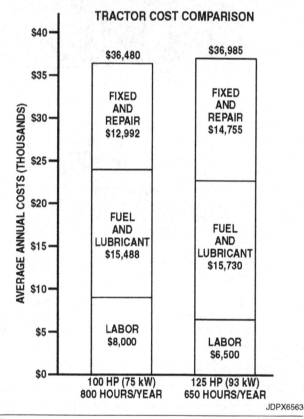

Fig. 4 — Comparison of Costs for Two Sizes of Tractors

Average costs for both the tractors (including labor) are shown in Fig. 4.

To summarize:

- Enough power is needed to get important jobs done within the allotted time.
- Keep costs down and do not buy excessively large tractors.
- Determine the value of the time saved by using the larger tractor.

Allowing for Expansion

When buying a tractor, there is another important factor to consider. Plan ahead and consider the possibility of the size of the operation increasing in the future. It is much better to buy a larger tractor than needed when trading than to have to trade up to a larger tractor in 3 or 4 years (Fig. 5).

The comparisons for the following examples are shown in Fig. 6, as options A and B.

Option A

Option A consists of buying a 125-horsepower (93-kW) two-wheel drive tractor with a list price of $80,000. It is used 400 hours per year for 4 years and then traded for a 160-horsepower (118-kW) front-wheel assist drive tractor because of the increase in acreage. The 160-horsepower tractor lists for $100,000 and is used for 400 hours per year.

Option B

Option B consists of buying a $100,000, 160-horsepower (118-kW) front-wheel-assist tractor in the beginning. It is used 300 hours a year for 4 years and then used for 400 hours a year when the operation expands.

Fig. 5 — It Is Better to Buy a Larger Tractor When Trading Than to Have to Trade for a Larger One in 3 or 4 Years

The process for making this mathematical comparison is complicated. The example is kept simple, but the principle is important to understand. The assumptions for our example are as follows:

- Diesel fuel is $4.00 per gallon ($1.06 per liter)
- Labor cost will not be included
- Number of acres involved will increase after 4 years

While labor cost should not be totally discounted, for simplicity sake it is left out of this equation. It can easily added to the cost by simply multiplying the labor rate by the number of hours each tractor is used.

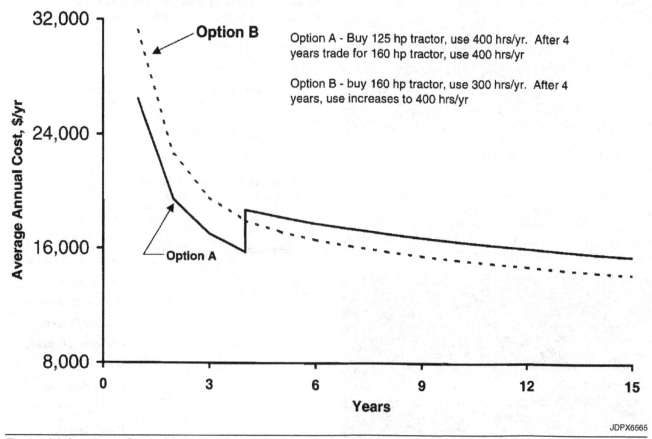

Fig. 6 — It Is Important to Buy the Proper Size Tractor if Expansion Is Anticipated

The average annual costs for both situations are shown in Fig. 6. Notice that average annual costs become less for Option B after 4 years and remains less even up to 15 years. This example shows the advantage of long range planning.

Selecting Machine Size

In buying a machine for farming, there are three important considerations:

- Selecting proper size for the power unit.
- Selecting sufficient capacity to complete tasks within the allotted time period.
- Making a decision that results in maximum net profit.

Matching the machine to a power unit has been discussed earlier in Chapter 4 and can be reviewed there. Determining capacity needed to get over a given number of acres within an allotted time period was also covered in Chapter 4. Two other important considerations are:

- Timeliness
- Alternative costs

Timeliness

Always consider the importance of timeliness when sizing machines (Fig. 7). Unfortunately, there is no easy way to calculate the answer for a timeliness factor.

Fig. 7 — Once a Crop Is Ready for Harvest, Field Losses Increase After That Point

The information in Table 3 is one way to estimate timeliness penalties in a machinery selection problem. Remember that staying ahead in important timeliness situations, such as planting corn (Fig. 8) or combining soybeans, means more bushels in the bin and a higher net profit.

Fig. 8 — Timely Planting Means Higher Yields

Timeliness Importance	Timeliness Loss, % Lost per Day of Delay	Timeliness Factor
Very High	1.0%	0.010
High	0.8%	0.008
Medium	0.6%	0.006
Low	0.4%	0.004
Very Low	0.2%	0.002

Table 3 — Estimating Timeliness Losses

These figures have been developed as a result of several research reports comparing the total crop produced and harvested for different time periods.

If a medium timeliness value (TV) of 0.6% is used, what is the timeliness loss (TL) for each day that 100 acres (40.5 hectares) of cotton goes unplanted (Fig. 8)? To estimate the timeliness loss, multiply the timeliness value (Table 4) by the number of acres times the number of days delayed.

100 acres (40.5 ha) x 0.006 x 1 day = 0.6 acres (0.24 ha) lost

These timeliness values assume a reduction in yield for a delay following the most favorable time period for the operation. The optimum time period might be only one day, or it might be several days, depending on the crop and the operation.

Timeliness Importance	Value, % per Day	Operation
Very High	1.0%	Corn Planting
		Sowing Oats
		Harvesting Soybeans
		Hay Operations
High	0.8%	Planting Soybeans
		Planting Peanuts
		Harvesting Peanuts
Medium	0.6%	Planting Cotton
		Harvesting Oats
Low	0.4%	Harvesting Wheat
		Harvesting Barley
		Harvesting Corn
		Harvesting Cotton

Table 4 — Timeliness Values Guide

Table 4 provides a guide for timeliness values (TV) for specific operations. It is only a guide and will not be the same for all situations. Check with local sources, such as the land-grant university, seed companies, etc., for timeliness values that apply to your location.

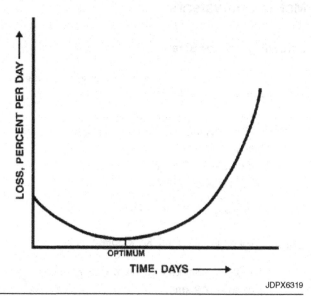

Fig. 9 — Timeliness Losses Can Occur Before and After the Optimum Period

In actual practice, timeliness values may apply to operations that start too soon as well as those that are delayed. Fig. 9 is one example of timeliness losses that applies to both situations. However, it is very complex and difficult to use. The straight-line timeliness loss in Fig. 7 is more commonly used. It is also the basis for Table 3 and Table 4.

A straight-line approach can be used to simulate sections of the parabolic curve in Fig. 9. For example, corn harvest losses might increase at the rate of 0.4% per day for the first 14 days following the optimum period. Then it might increase to 0.6% per day for the next 14 days. Finally, it could even jump to 1.0% per day after the second period.

The formula calculating time loss using the straight-line method is:

$$TL = \frac{Acres}{2} \times (days \div 1) \times TV$$

This formula assumes an equal amount is processed each day. Timeliness loss starts the first day following the optimum period.

Example: You have 500 acres (202.4 ha) of corn to plant (Fig. 10), of which 300 acres (121.5 ha) have been planted during the optimum period. Ten more days are required to plant the remaining 200 acres (80.9 ha). If the timeliness value (TV) is 1.0 % per day, what is the timeliness loss?

$$TL = \frac{200}{2} \times (10 \div 1) \times 0.1 = 11 \text{ acres}$$

The timeliness loss may be a loss of quality or a loss of quantity. In the case of a corn planting, it would be quantity. In harvesting corn, the quality loss would be a harvest loss that increases as the corn becomes drier due to stalks becoming brittle or lodged, or ears falling to ground.

Putting a Value on Timeliness

The application of the timeliness loss becomes more practical if the machinery cost is considered. A good example would be to compare three sizes of corn planters for planting 500 acres of corn.

Assume the optimum period for planting corn is May 1 to May 10 (10 calendar days). Corn planted anytime within this period will not have a yield reduction due to delayed planting. Starting on May 11, a yield loss of 1.0% per day is assumed. Long term weather records show an average of 5.5 working hours per day for this period.

Fig. 10 — Corn Planting is a Good Example of the Importance of Timeliness Values

Three sizes of planters (Fig. 10) will be considered, all equipped with pesticide attachments.

4-row 30 in. list price $15,000, 4.75 acres/hr

6-row 30 in. list price $18,000, 7.0 acres/hr

8-row 30 in. list price $23,800, 9.0 acres/hr

In the period of May 1 to 10, the following amounts would be planted:

4-row, 261 acres, leaving 239 acres, needs 10 more days

6-row, 385 acres, leaving 115 acres, needs 3 more days

8-row, 495 acres, leaving 5 acres, needs 1 more day

Timeliness losses would be:

4-row, $TL = \dfrac{239}{2} \times (10 \div 1) \times 0.1 = 13.1$ acres

6-row, $TL = \dfrac{115}{2} \times (3 \div 1) \times 0.1 = 2.3$ acres

8-row, $TL = \dfrac{5}{2} \times (1 \div 1) \times 0.1 = 0.05$ acre

Assume the corn yield averages 150 bushels per acre when planted from May 1 to May 10, and corn brings $6.00 per bushel, the timeliness losses (TL) would be:

4-row, 13.1 acres x 150 bu/acre x $6.00 = $11,790

6-row, 2.3 acres x 150 bu/acre x $6.00 = $2,070

8-row, 0.05 acres x 150 bu/acre x $6.00 = $45

Adding the cost of owning the planters to the timeliness costs can help determine the optimum size of the planter. To simplify the comparisons, only fixed costs will be considered. Operating costs per acre will be similar for all three planters. Assume all three planters will be owned for 10 years. Table 3, Chapter 6, shows average annual fixed costs for planters as 11.98% of initial list price for 10 years of ownership.

4-row, $15,000 x 0.1198 = $1,797 per year

6-row, $18,000 x 0.1198 = $2,156 per year

8-row, $23,000 x 0.1198 = $2,755 per year

Now, timeliness cost (TC) can be combined with fixed costs

Planter	TC	Fixed Costs	Total Costs
4-Row	$11,790	$1,797	$13,587
6-Row	$2,070	$2,156	$4,226
8-Row	$45	$2,755	$2,800

The costs for the 6-row and 8-row planters are less than half the cost of the 4-row planter when timeliness costs are considered.

This example shows how to add timeliness costs and machinery costs to help determine the optimum size. Combining timeliness costs and fixed costs is a quick and easy way to help determine the optimum size for a critical operation. While seeding and harvesting are normally considered to be the most critical operations, any operational delay could produce a "domino effect" on harvesting or planting.

Metric Equivalents

Selecting Tractor Size

The following formulas relate to information given earlier in this chapter in U.S. customary measurements.

Width, meters = $\dfrac{\text{hectares per hour} \times 10}{\text{Speed, km/h} \times \text{field efficiency \%}}$

Speed — 7.2 km/h
Capacity — 1.17 ha/hr
Field Efficiency — 80%

Width = $\dfrac{1.17 \times 10}{7.2 \times 0.8} = 2.03$ meters

Drawbar power, kW = $\dfrac{\text{Draft, kN} \times \text{Speed, km/h}}{3.6}$

Draft — 2.03m x 11.68 kN/m = 23.7 kN
Speed — 7.2 km/h

Drawbar kW = $\dfrac{23.7 \times 7.2}{3.6} = 47.4$ kW

Timeliness Losses

Metric equivalent to problem earlier in this chapter.

$TL = \dfrac{\text{hectares}}{2} \times (\text{days} \div 1) \times TV$

$TL = \dfrac{(80.9)}{2} \times (10 \div 1) \times 0.004 = 1.8$ ha

Summary

Planning tractor and machine purchases wisely is an important part of machinery management. Buying machines should also include sizing the machine for any future increases in the number of acres operated. Always buy the tractor or machine that is large enough to get the job done on time, even if financing is necessary.

Two important considerations to keep in mind when buying a tractor are:

- Provide for enough power to get all important field operations done on time.

- Provide for sufficient annual use so costs will be minimized.

Three factors to consider when buying a machine are:

- Selecting proper size for the power unit.

- Getting sufficient capacity to get needed work done within the allotted time period.

- Making a decision that results in maximum net profit.

Test Yourself

Questions

1. What are the three important factors to consider when deciding how large a machine to buy?

2. How much combining capacity is needed to combine 400 acres in 12 calendar days if you average 6 hours in the field per calendar day?

3. Which costs more per year to operate? How much more?

   a. A $60,000, 75-horsepower (56-kW) tractor used 600 hours per year

   b. A $80,000, 100-horsepower (75-kW) tractor used 450 hours per year

   Assume both are owned 8 years, diesel fuel costs $4.00 per gallon ($1.06 per liter), and labor is $10.00 per hour.

4. How would the answer for problem 3 change if labor costs $15.00 per hour?

5. If it takes 15 days to plant 600 acres (243 ha) of soybeans, what would the timeliness loss be if there is no reduction in yield for the first 5 days of the planting period?

6. Refer to problem 5; if the soybean yield is 40 bushels per acre (2.69 metric tons per ha) and soybeans sell for $6.00 per bushel ($220 per metric ton), what is the value of the yield reduction due to delayed planting?

7. How much more in annual fixed cost could be justified in problems 5 and 6 for a larger planter to complete the planting within the optimum period? Convert the annual fixed cost to an additional list price for the planter, assuming a 10-year ownership.

8. Think of a critical planting or harvesting operation in your area and see what the timeliness losses would be for different numbers of days to complete the operation. Make your own assumptions for crop yield, value, and length of the optimum period. Then determine the optimum size of the machine.

Calculating Custom Work Costs

Introduction

Fig. 1 — For Some Operations, It May Be Better to Use a Custom Operator

Hiring custom operators is one important alternative to owning machinery. In some cases, using custom operators completes the work faster, provides the least-cost method, and does not require the capital needed for owning a machine (Fig. 1). In other cases, doing additional custom work can help a farm operator justify ownership costs.

When considering hiring a custom operator, do not forget timeliness considerations. Waiting for a custom operator to arrive is expensive in terms of timeliness if it means not getting crops planted or harvested at the optimum time. Always consider timeliness in making a decision to use either a custom operator or an alternative method.

Chapter Objectives

- Calculate custom costs
- Compare custom costs to cost of ownership
- Determine number of acres of custom work needed to break even

Determining and Comparing Costs

Fig. 2 — For Small Annual Use, a Custom Operator May Provide the Least-Cost Method

Determining when to use a custom operator is one of the most important decisions made in machinery management (Fig. 2). There is a rather simple formula for calculating the break-even point for owning machinery versus the cost of custom work.

The formula for calculating the break-even point is:

$$\text{Break-Even Point} = \frac{\text{Average Annual Fixed Cost}}{\text{Custom Rate/Unit Area} - \text{Operating Costs/Unit Area}}$$

As explained in Chapter 6, fixed costs are incurred by owning a machine, regardless of annual use. In the case of the custom operator, these fixed costs must be obtained from the custom charge. Operating costs, which include costs for fuel, lubricants, labor, and repairs, must also be obtained from the custom charge.

A custom operator receives additional income from each acre (or hectare) to pay off the annual fixed costs. The following example illustrates the relationship between fixed costs, operating costs, and custom charges.

Suppose annual fixed costs are $2,500 per year to own a pull-type windrower, and total operating costs are $3.50 per acre for a tractor, labor, fuel, lubricants, and repairs. Assume a custom operator contracts to use the windrower on 300 acres per year and charges $12.00 per acre. What are the annual fixed cost, operating cost, and income for custom work per year?

Annual fixed cost = $2,500/year

Annual operating cost = $3.50/acre × 300 acres = $1,050/year

Income for custom work = $12.00/acre x 300 acres = $3,600/year

The difference between the income for custom work and operating costs is $2,550 per year ($3,600/year − $1,050/year). The difference of $2,550 per year is the amount the custom operator has left, after paying operating costs, to pay fixed costs and for profit. In this case, the difference of $2,550 covers the fixed costs of $2,500 and leaves a profit of just $50. With just $50 profit, the custom operator is just barely breaking even.

Now determine the number of acres needed to break even. (See Metric Equivalents for solution in metric units.) The fixed costs are $2,500 per year. The difference between income for custom work and operating costs is $2,550 per year or $8.50 per acre ($2,550/year divided by 300 acre/year). Using the formula to calculate the break-even point gives:

Break-even point, acres =
$$\frac{\$2,500/\text{year}}{\$12.00/\text{acre} - \$3.50/\text{acre}} = 294 \text{ acres}$$

Thus, the formula verifies that if the custom operator of the pull-type windrower is able to contract for 300 acres per year, the operation just barely breaks even. The profit is $50 per year. To increase profit, more custom work would be needed. The effect of increasing the number of acres per year on costs and profit is shown in Fig. 3.

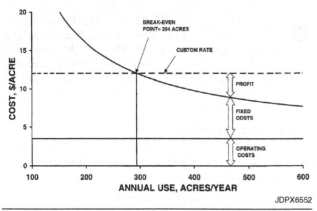

Fig. 3 — Example Problem Pull-Type Windrower Shows Ownership Is Justified for Annual Use of 294 Acres or More

To use this method for each situation, know or be able to determine the following:

- Annual fixed costs
- Average operating costs

Let's look at each of these costs in detail.

Determining Annual Fixed Cost

The annual fixed cost can be determined by knowing the cost of a machine and length of ownership. For example, if a chisel plow is to be owned for 10 years, the average annual fixed cost is 11.98% of the list price, as shown in Table 3, Chapter 6, and shown here as Table 1.

Age, Years	Tractors and Combines	Forage Harvesters	All Others
1	33.07%	35.75%	34.41%
2	22.52%	24.81%	23.68%
3	18.76%	20.70%	19.75%
4	16.71%	18.33%	17.55%
5	15.36%	16.69%	16.06%
6	14.36%	15.42%	14.93%
7	13.56%	14.39%	14.01%
8	12.90%	13.51%	13.24%
9	12.33%	12.75%	12.57%
10	11.82%	12.07%	**11.98%**
11	11.37%	11.46%	11.45%
12	10.96%	10.91%	10.96%
13	10.58%	10.41%	10.52%
14	10.23%	9.94%	10.11%
15	9.91%	9.51%	9.72%

Table 1 — Average Annual Fixed Cost as a Percentage of Original List Price

The best bet is to be realistic in setting the length of ownership when estimating the average annual fixed cost. It would be a mistake to justify ownership on the basis of a 10-year life and then trade in a lesser period of time, such as 4 years.

The effect of length of ownership on the break-even point for the previous example of a windrower is illustrated in Fig. 4.

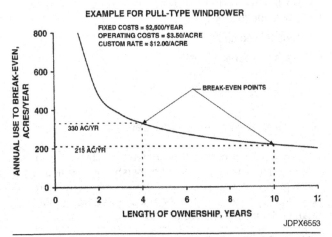

Fig. 4 — To Reach the Break-Even Point, the Length of Ownership Decreases as Annual Use Increases

Determining Average Operating Costs

Operating costs can be estimated from records or from the earlier chapters in this book. Remember that operating costs include:

- Fuel
- Lubricants
- Labor
- Repairs

Determining Total Costs

Below is another example of the break-even point, using a large square baler. Assumptions include:

- List price for the baler is $65,000
- Capacity is 15 tons (13.6 MT) per hour
- Custom rate is $15.00 per ton ($16.54 per MT) (Custom rates will vary even for the same crop, depending on area, crop yield, and supporting services provided, such as hauling.)
- Tractor costs are $25.00 an hour
- Labor is $10.00 an hour
- For a crop yield of 1.5 tons per acre (3.36 MT per hectare), capacity is 10 acres (4.05 hectares) per hour
- Twine cost is $1.00 per ton ($1.10 per MT)

Assume the baler will be traded at 600 hours. Repair costs in this case will be $3,735, or $6.22 per hour (Chapter 8).

Step 1 — Determine Average Annual Fixed Cost

Assuming a 10-year life for the baler, 11.98% (Fig. 4) of the new cost of the baler is the average annual cost for depreciation, taxes, shelter, insurance, and interest (fixed costs):

Average Annual Fixed Cost =
11.98% of $65,000 = 0.1198 x $65,000 = $7,787 per year

Using the above capacity and custom rate, the custom rate is $22.50 per acre ($15 per ton times 1.5 tons per acre).

Step 2 — Determine Operating Costs

Operating costs include cost for twine at $1.00 per ton ($1.10 per MT). Twine cost is $1.50 per acre ($1.00/ton x 1.5 tons/acre) and in metric units twine cost is $3.70 per hectare ($1.10/MT x 3.36 MT/ha). Twine cost is $15.00 per hour ($1.50/acre x 10 acres/hr) or in metric units ($3.70/ha x 4.05 ha/hr).

Now add all the operating costs. The total operating costs for the baler are:

Tractor = $25.00 per hour

Labor = $10.00 per hour

Repairs = $6.22 per hour

Twine cost = $15.00 per hour

Total = $56.22 per hour

Total operating cost is $5.62 per acre ($56.22/hr divided by 10 acres/hr). The difference between the custom cost and operating cost is therefore $16.88 per acre ($22.50/acre – $5.62/acre).

 Operating cost is $13.88 per hectare ($56.22/hr divided by 4.05 ha/hr). The custom rate is $55.57 per hectare ($16.54/MT x 3.36 MT/ha). The difference between the custom rate of $55.57 per hectare and the operating cost of $13.88 per hectare is $41.69 per hectare.

Step 3 — Determine Break-Even Point

Now calculate the break-even-point. The formula is:

$$\text{Break-Even Point, Acres} = \frac{\text{Average Annual Fixed Cost}}{(\text{Custom Rate/Acre} - \text{Operating Costs/Acre})}$$

Inserting the various costs for the baler into the equation we get:

$$\text{Break-Even Point, Acres} = \frac{\$1,787/\text{yr}}{\$22.50/\text{acre} - \$5.62/\text{acre}} = 461 \text{ acres}$$

In this case, 461 acres or more are needed for ownership to be less costly than hiring a custom operator (Fig. 5).

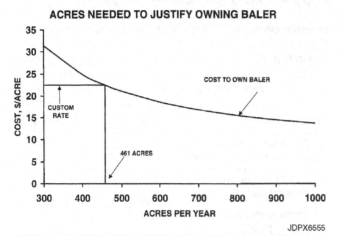

Fig. 5 — Owning Machinery Can Be Justified by Some Custom Work

Now an interesting question arises. Suppose you have only 400 acres (162 hectares) per year to bale, but you are willing to accept the higher cost of ownership so you can bale when the crop is ready (Fig. 6). What is the additional cost to own the baler instead of using a custom operator?

Fig. 6 — Some Operators Prefer Owning Machinery to Ensure a Timely Harvest

For 400 acres at $22.50 per acre, $9,000 per year would be spent for custom work. To own the baler, fixed cost is $7,787 per year. Operating cost is $5.62 per acre, or $2,248 per year (400 acres x $5.62/acre). This means the total cost is $10,035 per year ($7,787 + $2,248). Therefore, the extra cost to own the baler in this case would be $1,035 per year ($10,035/yr − $9,000/yr) or $2.59 per acre ($6.39 per hectare).

Doing Custom Work to Help Justify Ownership Cost

Another question often comes up regarding custom work and owning. Some farmers like to do custom work in addition to their own work in order to justify ownership (Fig. 8). By determining the difference between the total work on your farm and the break-even point, the additional acres of custom work needed to justify ownership can be determined.

Fig. 7 — Some Machinery Owners Do Only Custom Work

In the previous example, the difference between the break-even point of 461 acres (187 ha) and 400 acres (162 ha) on your farm is 61 acres (25 ha). Therefore, 61 acres (25 ha) of custom work is needed to justify ownership.

Establishing a Rate for Custom Work

In addition to ownership and operating costs, establishing a rate to charge for custom work depends on several variables, including:

- Size of fields
- Travel distance from home base or last job
- Difficulties due to crop or field conditions
- Labor costs
- Profit-and-risk margins

The approach to setting a custom rate is basically a matter of accurately estimating costs, then adding a margin for profit and risk. An adjustment for the specific situation is the final step.

Consider the situation of a custom business expected to return a profit. In this case, the procedure consists of three steps:

1. Estimating annual use
2. Determining the basic cost
3. Adding a margin to cover risk and profit

As an example, suppose you want to run a custom wheat-combining operation, and plan to buy a $220,000 combine that can harvest 8 acres (3.24 ha) of wheat per hour in average conditions.

Labor is $10.00 per hour. Total annual use is estimated at 200 hours per year. The combine has a 160-horsepower (120-kW) diesel engine that averages 8 gallons (30.3 liters) of fuel per hour. The fuel costs $4.00 per gallon ($1.06 per liter).

Step 1 — Determine Annual Use

Annual use is already estimated at 200 hours per year.

Step 2 — Determine Basic Costs

From information in Chapter 10, we determined that the combine could be owned 10 years or 3,000 hours, whichever comes first. In this case, assume the combine would be owned 10 years. The fixed costs plus repair costs are $0.677 per hour per $1,000 of list price (Table 3, Appendix).

The fixed cost plus repair cost for the $220,000 combine are:

$220,000 divided by $1,000 x $0.677 = $148.94 per hour

Fuel cost would be:

8 gallons per hour x $4.00 per gallon = $32.00 per hour

30.3 liters per hour x $1.06 per liter = $32.12 per hour

Add 10% of fuel cost for lubricant cost to obtain total fuel and lubricant cost:

$32.00 + (0.10 x $32.00) = $35.20 per hour

SUMMARY OF COSTS:

Fixed cost plus repair cost = $101.55 per hour

Fuel and lubricant cost = $35.20 per hour

Labor = $10.00 per hour

Total = $146.75 per hour

At 8 acres (3.24 ha) per hour, the cost would be:

$146.75 divided by 8 = $18.34 per acre

$146.75 divided by 3.24 = $45.29 per hectare

The basic total cost is $18.34 per acre ($45.29/ha), including all fixed and repair costs, and costs for fuel, lubricants, and labor.

Step 3 — Add the Profit-and-Risk Margin

Determining the profit-and-risk margin is a tricky procedure (Table 2). This margin may range from 20% to 60%, depending on the complexity of a machine. A complex machine like a combine has a much greater chance for a breakdown than a simpler tool like a disk harrow.

Typical Operation	Average Custom Charges
Light Tillage (Very Little Risk, Favorable Field Conditions)	Cost plus 20% to 30%
Plowing or Chiseling (Average Field Conditions)	Cost plus 30% to 40%
Medium to Heavy Disking (Average Risk)	
Hay and Silage Operation (Heavy Risks)	Cost plus 40% to 50%
Combining (Severe Field Conditions)	

Table 2 — Custom Work Charges

In this example, a profit-and-risk margin of 40% might be appropriate for the combine.

The custom rate charge would be calculated as follows:

$18.34 + (0.40 x $18.34) = $25.68 per acre

$45.29 + (0.40 x $45.29) = $63.41 per hectare

The custom rate charge, therefore, includes fixed and repair costs, costs for fuel and lubricants and labor, plus a charge to cover risk and profit (Fig. 8). At this point, a likely custom rate would be set at $26.00 per acre ($64.00/ha). Adding charges for equipment and labor to support the operation, such as trucks, may result in a final custom charge closer to $33.43 per acre ($81.20 per hectare) for this example.

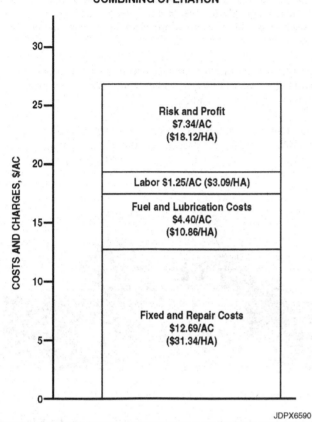

Fig. 8 — Custom Rates Should Include All Costs Plus a Charge for Risks and Profit

This example illustrates just one procedure that can be used for setting custom rates. The markup for risk and profit depends on individual situations. If there is a lot of competition, the margin may be quite small. If there is little competition and the operation is a fairly new method, the margin may be quite high. Information in Table 2 can be used as a guide to help compute margin for custom rates.

Calculating Amount of Custom Work Needed to Justify Ownership

Suppose you are thinking of buying a baler to do custom work and want to know how many bales per year it would take to break even. Assume the situation is as follows:

- Baler price = $16,000
- Tractor cost = $15.00 per hour
- Labor cost = $10.00 per hour
- Custom charge = $0.60 per bale
- Overall capacity = 100 bales per hour
- Wire cost = $0.15 per bale

First, determine the operating cost per bale. For repair costs, use a half-life for the baler (Chapter 8, Table 2). For a small square baler, total repair costs from when the baler is new to its half-life (1,250 hours) would be 21.5% of the $16,000 list price or $3,440. Dividing $3,440 by 1,250 hours gives us an estimated repair cost of $2.75 per hour.

Operating cost would be:

Tractor = $15.00/hr

Labor = $10.00/hr

Repairs = $2.75/hr

Total = $27.75/hr

$$\frac{\$27.75/hour}{100\ bales/hour} = \$0.28/bale$$

Add wire cost of $0.15 per bale for a total operating cost of $0.43 per bale.

Now we need to determine the average fixed cost using information from Chapter 6. For a serious custom baling operation, ownership could range from 4 to 6 years. For this example, assume a 6-year ownership period.

Average annual fixed cost for 6 years would be 14.93% times the list price (Chapter 6, Table 3). The average fixed cost would be:

$$0.1493 \times \$16,000 = \$2,389/yr$$

We are now ready to use the break-even formula, which is:

Break-Even Point, Bales =

$$\frac{Average\ Annual\ Fixed\ Cost}{Custom\ Rate/Bale - Operating\ Cost/Bale}$$

The number of bales per year to break even is:

$$\frac{\$2,389/yr}{\$0.60/bale - \$0.43/bale} = 13,850\ bales/yr$$

At 100 bales per hour, the break-even point in hours of use would be:

$$\frac{13,850\ bales/year}{100\ bales/hour} = 138\ hours/year$$

This means the average annual fixed cost and the operating costs would be covered by the custom charge when the baler is used to make 13,850 bales each year or when the baler is used 138 hours per year (Table 3).

Hours per Year	Bales per Year	Profit, $/Year
0	0	−2,389
25	2,500	−1,958
50	5,000	−1,526
100	10,000	−664
138	13,850	Break-Even
150	15,000	198
200	20,000	1,061
250	25,000	1,923
300	30,000	2,786
400	40,000	4,510

Table 3 — Custom Operator Makes Profit Only After Break-Even Point

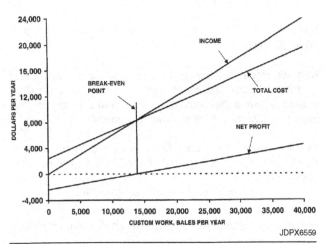

Fig. 9 — The Amount of Custom Work Affects Income, Costs, and Profits

Figure 9 shows the break-even point, costs, and profit for different amounts of custom work. If the amount of custom work is less than 13,850 bales per year, profit is negative, which means the custom operator loses money. If the amount of custom work is more than 13,850 bales per year, the custom operator is able to make a profit and cover the risks involved. The amount for profit and risk after the break-even point would be $0.17 per bale (custom charge of $0.60/bale minus the operating cost of $0.43/bale).

Other important factors that should be taken into consideration in setting the rate for custom work are travel distance from the home base, field conditions, crop yields, and other factors that might affect daily capacity. It is better to invest money elsewhere than to do custom work at such a low rate that expenses are barely recovered.

From this example, an important conclusion can be made. When considering custom work, the greater the volume, the greater the profit.

Adding Custom Work to Reduce Costs

This chapter has dealt primarily with the procedure for establishing custom rates for a business. A similar procedure may be used to establish charging rates for part-time custom work.

Part-time custom work can be beneficial to both parties. If equipment is not fully utilized, fixed costs are probably higher than they should be.

For example, suppose a tractor and plow used are in good condition. The tractor is used 500 hours a year and the plow 80 hours a year. A neighbor wants 120 acres plowed on a custom hire basis.

Fig. 10 — Custom Work Can Benefit Both Parties

The 10-year-old 100-hp tractor originally listed for $45,000 and the 8-year-old plow with 3.4 acres per hour capacity had a list price of $8,000.

What should be charged for the custom operation? How much will it help to offset ownership costs?

The tractor's fixed cost plus repairs =

$\left(\dfrac{\$45,000}{1,000}\right)$ x $0.271 (Table 1, Appendix) = $12.20/hr

Fuel and lube costs =
100 hp x 0.044 gal/hp x $4.00 hr/gal x 1.1 = $19.36//hr

Labor = $10.00/hr

Total tractor costs =
$12.20/hr + $19.36/hr + $10.00/hr = $41.56/hr

Plow costs:

Fixed costs plus repairs =

$\left(\dfrac{\$8,000}{1,000}\right)$ x $1.857 (Table 6, Appendix) = $14.86/hr

Total tractor and plow costs =
$41.56/hr + $14.86/hr = $56.42/hr

The plowing operation costs $56.42 per hour. Add 20% of cost for profit and to cover risk.

$56.42 + (20% of $56.42) = $67.70 per hour. Divide $68.00 per hour ($67.70 rounded) by 3.4 acres per hour and charge $20.00 per acre for the plowing.

The total income would be 120 acres x $20.00 or $2,400.

Plowing 120 acres at 3.4 acres per hour would take 35.3 hours (120 acres/3.4 acres per hour). Out-of-pocket costs would include costs for fuel, lubricant, and labor, which would be:

35.3 hours x ($19.36/hr + $10.00/hr) = $1,036.41

The amount to cover fixed and repair costs would be $12.20 per hour for the tractor and $14.86 per hour for the plow, for a total of $27.06 per hour. For 35.3 hours of work, the amount to cover fixed and repair costs would be:

35.5 hours x $27.06/hr = $960.63

Therefore, the total costs would be $1,997.04 ($1036.41 + $955.00). Profits would equal the income of $2,400.00 minus the costs of $1,997.04 which equals $402.96.

Metric Equivalents

Determining and Comparing Cost

The following example, in metrics, is roughly equivalent to the example shown in U.S. customary units earlier in this chapter.

Custom operator charges = $30.00 per hectare

Cost per year to own pull-type windrower = $2,500

Operating cost for windrower (tractor, labor, fuel, oil, lubricants, repairs) = $8.65/hectare

If the windrower has 120 hectares of use per year, the difference between custom costs ($30.00 x 120 = $3,600) and operating costs ($8.65 x 120 = $1,038) would be the real cost for the custom operator. In this case, the real cost would be $2,562, which almost exactly equals the average annual fixed costs.

Because a cost of $21.35 per hectare is made, 117 hectares ($2,500/$21.35) is needed to justify ownership. Using the formula to calculate the break-even point gives the same answer.

Break-Even Hectares $= \dfrac{\$2,500}{\$30.00 - \$8.65} =$ 117 hectares

Summary

A custom operator may be the least-cost choice compared to owning a machine in some cases. One advantage of hiring a custom operator is that less money is tied up in ownership of machinery. One disadvantage may be the lack of timeliness.

The formula used for calculating the number of acres or hectares needed per year to justify ownership of a machine is:

$$\frac{\text{Average Annual Fixed Costs}}{\text{Custom Rate per Area} - \text{Operating Cost per Area}}$$

To estimate the costs involved in a custom operation, there are three steps:

- Estimating annual use
- Determining the basic cost
- Adding a margin for profit and risk

Fig. 11 — An Accurate Cost Analysis Is Essential to Custom Work

An accurate cost analysis is the key to establishing custom charges (Fig. 11). If doing custom work will not return a profit and help offset ownership costs, then it is unwise to be in the business. A common mistake is to justify purchase of a machine on the basis of doing custom work and then setting your custom charge at a rate that barely returns your operating costs.

Test Yourself

Questions

1. When would you use a custom operator instead of buying a machine? (Give two examples.)

2. When would you buy a machine rather than using a custom operator? (Give two examples.)

3. How many acres per year would you need to justify ownership of a self-propelled combine if average annual fixed cost is $12,000, the custom rate is $20.00 per acre, and your operating costs are $4.00 per acre?

4. Suppose you have 600 acres of work per year. Could you justify the combine described in problem 3? If not, what would be the added cost per acre to own rather than use a custom operator?

5. You own a self-propelled windrower that has a total cost of $40.00 per hour. Capacity is 5 acres per hour. What would you suggest as a charge for custom work?

6. How many hectares of wheat must you combine per year to break even for doing custom work if you charge $50.00 per hectare, capacity is 3.25 hectares per hour, repair costs are $3.00 per hour of use, labor is $10.00 per hour, fuel and lubricant costs are $8.00 per hour, the combine costs $100,000; you use it 300 hours per year, and will trade in 6 years.?

Decision Time — Selecting the Best Alternative

13

Introduction

Fig. 1 — Continually Evaluate Alternatives

Good machinery management is a constant process of evaluating alternatives (Fig. 1). Limiting the options for consideration is a common mistake. Personal prejudice, pride, and sometimes available capital can narrow alternatives and preclude what may well be the best answer. The overriding goal is to keep costs at a minimum while achieving profit goals.

The best first step is to stay informed. What are the local rental rates? Are long-term leases available? Is good service available? How much would it cost to repair the current machine? How much would it cost to trade? How large a machine do you really need? How much can you safely borrow?

So far, the focus has been on sizing and matching equipment, cost of ownership, and decisions on when to trade, when to use a custom operator, and when to do custom work. This chapter brings it all together in the application of knowledge to identify and select the best alternative.

Chapter Objective

- Compare alternatives of repairing, trading and renting or leasing.

What Are the Alternatives?

If your tractor is 12 years old, has 6,600 hours, and needs extensive repairs, it's decision time. The alternatives include:

- Rent
- Lease
- Repair and keep on owning
- Buy a new tractor of the same size
- Sell it and make do with the other tractors
- Buy a new tractor of a different size
- Buy a used tractor

Cost of owning and operating the equipment is the final criterion for decision-making, providing it meets the operational needs.

Following a discussion of renting and leasing, a method will be presented that can be used for comparison of all of the various alternatives.

Renting

Renting is generally considered as a short-term solution that only defers a decision. Below are several situations where renting might be considered as the best alternative.

- Low annual use. Subsoiling or deep ripping an area too small to justify ownership.
- Deferring capital expenditures. If you are in a cash or credit crunch, renting can be a short-term solution.

- Periods of uncertainty. Changes in the farming operation, either enterprises or size of the operation are two examples.
- Timeliness. If, for example, fall tillage could not be completed as planned, you may need more power to get caught up in the spring.
- Competitive cost opportunity. Local rental costs may be lower than cost of ownership, particularly if it helps defer a purchase. It may be a case where timeliness justifies renting more capacity than you can justify owning.
- More reliability. Usually rental machines either are new or have very little use.

Renting has two major drawbacks. Rental rates cannot be guaranteed on a long-term basis, usually only a year at a time (Fig. 2). Second, the rental machine needs to be available when you need it.

1. TERM; RENTAL

Lessor hereby leases the above units(s) to Less
advance as follows (Check applicable square.):

☐ MONTHLY, QUARTERLY, OR SEMI-ANNUAL
for a total of ______________ rental payments.
(In the case of monthly rental payments, the initial

☐ ANNUAL. A rental payment of $ ______________
annually thereafter on the anniversary date of the

2. ASSIGNMENT BY LESSOR

Lessee agrees that Lessor may assign this lease a
ment Lessee hereby waives notice. Lessee's obliga
of any obligation hereunder or other liability of Le
the "Owner." Lessee will not sublease the unit(s)

3. EXTENSION

JDPX6302

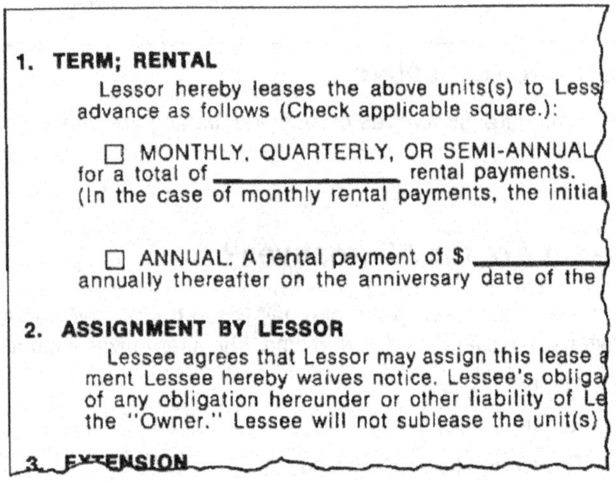

Fig. 2 — The Term "Rent" Usually Applies to a Period of Time Lasting Less Than 12 Months

Leasing

JDPX6303

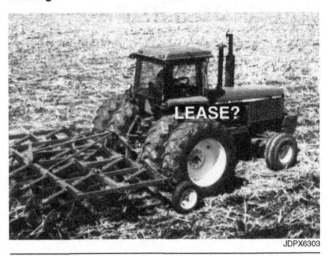

Fig. 3 — Leasing Is a Long-Term Commitment

There are some situations where leasing may be the best method of obtaining machinery (Fig. 3). Some of the more important reasons include:

- There is a shortage of capital, such as when expanding the business.
- A short-term opportunity investment in the business occurs that will yield a greater return on capital than financing costs for purchasing equipment.
- The rate of return of the business is considerably higher than interest rates on borrowed capital.
- The future is uncertain and it is preferred to defer long-term commitments. However, the commitment to lease is the same as signing a sales contract as far as financial obligations are concerned.
- There is a definite reliability and operational advantage in having new, late-model equipment, such as with leasing.
- There is uncertainty as to what size or model is needed in your farming situation.
- The owner is phasing out of business.

Leasing opportunities come and go, but have not remained a very viable alternative. In some cases, corporate farms may be able to justify leasing of some equipment.

Costs of renting or leasing need to be determined on a per-acre basis (Fig. 4).

Fig. 4 — Figure Rental Costs on a Per-Acre Basis

For example, you may be able to rent a wheat combine that has a list price of $140,000 for $120.00 an hour. If the combine averages 7.27 acres per hour, then costs might look as follows:

Rental charge	$120.00 an hour
Labor	$10.00 an hour
Fuel and Lube	$11.00 an hour
Total costs	$141.00 an hour
At 7.27 acres/hour, costs =	$19.39 an acre

A consideration may be made for income tax deductions. A proper rental agreement allows full deductions of rental costs. If you rented the machine above and were in the 28% tax bracket, the cost would, in effect, be discounted 28%:

$19.39 less 28% of $19.39 would be: $13.96 per acre.

Comparing Alternatives

It is difficult to compare more than two alternatives at the same time without getting into complex procedures.

The cost section of this text provided the resources and methodology for cost determinations. That method is satisfactory for comparing alternatives involving new purchases for specific periods of ownership.

Use the form in Fig. 6 to compare up to four alternatives at the same time. This is not limited to new purchases. It is well suited for comparing alternatives that include "repair and keep" and buying a used machine. This form is also provided in the Appendix.

This shortcut method is designed to provide a comparative cost analysis that includes your own input information. Fixed costs, including interest and depreciation, can be based on actual figures. The only estimates needed are those for projected repair costs and, occasionally, "as-is" values.

Operating costs of fuel and lubricants (Fig. 5), and labor are optional inputs. However, if they are used for one option in a comparison, they should be used for all options. These costs should normally be included, particularly when using a custom operator is one of the options.

Fig. 5 — Fuel and Lubricant Costs Are the Same Whether the Machine Is Leased or Owned

Charges for housing, insurance, and taxes are also included because they are fixed costs that could have a significant influence on the final decision. These costs are estimated at 4% of the average value.

The advantage of this procedure is to include income tax deductions for the machinery costs. It also provides a place to insert an estimate of the timeliness costs due to machine reliability or delays in operations.

NOTE: The Worksheet for Comparing Alternatives form shown in Fig. 6 is set up to always use Option 1 for the "repair and keep" alternative, due to line i.

Worksheet For Comparing Alternatives

Options ➝	1	2	3	4
Machine:	Repair and Keep Using			
a. Age at Purchase, Years				
b. Current Age, Years				
c. Ending Age, Years				
d. Use, Acres/Year				
e. Use, Hectares/Year				
f. Original List Price				
g. "As-Is" Value (Chapter 6, Table 1)				
h. "Cash Boot" Difference				
i. Capital Value (Line g of Option 1 + Line h of Each Option)				
j. Ending "As-Is" Value				
k. Loss in Capital Value				
l. Immediate Repairs				
m. Future Repairs				
n. Finance Charges on Investment Capital				
o. Taxes, Shelter, and Insurance Costs				
p. Total (Lines k + l + m + n + o)				
q. Annual Costs, $/Year (Line p/[Line c − Line b])				
r. Adjustments, $/Year				
Labor, $/Year				
Fuel and Lubricant, $/Year				
Age Penalty, $/Year				
Income Tax Credit[a], $/Year				
s. Total Cost, $/Year (Line q + r)				
t. Total Cost, $/Acre (Line s/d)				

a. For income tax credit, use the annual average loss in capital value for the depreciation of line i, plus lines l, m, n, and r (labor, fuel, and lubrication costs). Depreciable portion of line i is normally the "cash boot" difference.

Fig. 6 — Worksheet for Comparing Alternatives

The following examples will show how to use the form:

Comparing Two Alternatives

A wheat combine is 8 years old and has reached the point where something must be done, since it needs $7,000 in repairs. It has been used for 800 acres and 110 hours a year since it was purchased 8 years ago. Two options will be considered (Fig. 7).

Fig. 7 — Repair the Old Combine or Trade It for a New One?

Option 1: Repair the current combine and own it for 10 more years.

Option 2: Trade for a new combine with a list price of $240,000 and own it for 10 years.

When using this form (Fig. 8) to make comparisons, it is advisable to use the same period of ownership for each option.

Assumptions: Both combines have the same capacity of 7.3 acres per hour. They use 1 gallon of diesel fuel per acre at $4.00 a gallon. Labor is $10.00 an hour.

Remaining values and repair cost projections are based on formulas from Section II.

Finance charge is based on 11% of the remaining machine values per year, and is applied to the capital value of machine and immediate repair cost.

An income tax bracket of 28% is used, with no assumptions for state income tax deductions. Tax deductions for interest charges apply only to cost of immediate repairs and finance charge on capital invested.

Machinery is fully depreciated over a 7-year period, for income tax purposes only.

In working through this example, it is more important to learn and understand how to use the form than to draw conclusions from the results. The following explanations of the steps and the completed form (Fig. 8) for the example should help.

Lines a through f. Self-explanatory.

Line g. "As-is" value for Option 1, the old combine, is $40,841. The "as-is" value for the new combine is estimated to be 85% of its list price, or $204,000 (0.85 x $240,000).

Line h. "Cash boot" difference is the amount of capital required to trade for a new combine. For Option 2, the "cash boot" difference is $163,159 ($204,000 – $40,841).

Line i. Capital value is the total amount invested. For Option 2, the capital value is $204,000, which is equal to the trade-in value of the old combine ($40,841) plus the "cash boot" difference for the new combine ($163,159). In this case, the capital value is equal to the purchase price of the new combine.

Line j. Ending "as-is" value is estimated using line f and a value from Table 1, Chapter 6.

Line k. Loss in capital value is the depreciation over the ownership period. In Option 2, depreciation is equal to $118,440, the capital value ($204,000) minus the ending "as-is" value ($85,560).

Line l. Immediate repair cost is the best estimate of needed repair costs. For Option 1, the combine will be repaired and owned for 10 more years; it is estimated that $7,000 is needed for immediate repairs.

Line m. Future repair costs are estimated using formulas from Chapter 8.

Line n. Finance charges apply to the value of the combine plus immediate repair costs. Using a shortcut method, the finance charge is equal to the average value times the interest rate times the number of years.

For Option 1, finance charge would be:

$$\left(\frac{(\$40{,}841 + \$21{,}998)}{2} + \$7{,}000\right) \times 0.11 \times 10 = \$42{,}261$$

For Option 2, finance charge would be:

$$\left(\frac{(\$204{,}000 + \$85{,}560)}{2}\right) \times 0.11 \times 10 = \$159{,}258$$

Line o. Taxes, shelter, and insurance cost is equal to 4% of the average value times the number of years:

$$0.04 \times \left(\frac{\text{Line i} + \text{Line j}}{2}\right) \times (\text{Line c} - \text{Line b})$$

Line p. Total cost equals the sum of lines k, l, m, n, and o.

Line q. Average annual cost is equal to:

$$\frac{\text{Line p}}{\text{Line c} - \text{Line b}}$$

Line r. Adjustments to annual costs for fuel, lubricants, and labor are optional. However, if these costs are included in one option, they should be included in all options.

An age penalty is any loss in crop quality or quantity expected because of the age of a machine. In Option 1, an age penalty of $1,500 per year is estimated for the old combine due to expected high harvest losses.

Other adjustments might include deductions for income tax credits. Credits are applied to lines k, l, m, n, and o, and to costs for fuel, lubricants, and labor in line r. A marginal income tax rate of 28% is assumed. For Option 1, it is assumed depreciation has already been used for income tax purposes and, therefore, the deduction for Option 1 is:

$$0.28 \times \left(\frac{\$7,000 + \$13,670 + \$42,261 + \$12,568}{10} + [\$1,100 + \$3,520] \right)$$

$= \$3,407.57/\text{Year}$

Lines s and t. Total cost is $10,087 per year or $12.61 per acre for Option 1, repair and keep, using the old combine. For the new combine, the cost is $15,842 per year or $19.80 per acre. The savings for repair and continuing to use the old combine is $5,755 per year or $7.19 per acre. This comparison illustrates, as is often the case, that it is less expensive to repair and keep using an old machine. Of course there are limits to how long a machine should be used. Reliability may become a major factor in using the old machine for a total of 20 years, making it more economical to trade for a new machine. Reliability is not accounted for in the comparison.

Worksheet For Comparing Alternatives

Options ➜	1	2	3	4
Machine: Wheat Combine	Repair and Keep Using	Trade for New		
a. Age at Purchase, Years	0	0		
b. Current Age, Years	8	0		
c. Ending Age, Years	18	10		
d. Use, Acres/Year	800	800		
e. Use, Hectares/Year	110	110		
f. Original List Price	$100,000	$204,000		
g. "As-Is" Value (Chapter 6, Table 1)	$40,841	$204,000		
h. "Cash Boot" Difference	----	$163,159		
i. Capital Value (Line g of Option 1 + Line h of Each Option)	$40.841	$204,000		
j. Ending "As-Is" Value	$21,998	$85,560		
k. Loss in Capital Value	$18,843	$118,440		
l. Immediate Repairs	$7,000	$0		
m. Future Repairs	$13,670	$6,810		
n. Finance Charges on Investment Capital	$42,261	$159,258		
o. Taxes, Shelter, and Insurance Costs	$12,658	$57,910		
p. Total (Lines k + l + m + n + o)	$94,432	$342,418		
q. Annual Costs, $/Year (Line p/[Line c − Line b])	$9,443	$34,242		
r. Adjustments, $/Year				
Labor, $/Year	$1,100	$1,100		
Fuel and Lubricant, $/Year	$3,520	$3,520		
Age Penalty, $/Year	$1,500	$0		
Income Tax Credit[a], $/Year	($3407)	($6,161)		
s. Total Cost, $/Year (Line q + r)	$12,956	$32,701		
t. Total Cost, $/Acre (Line s/d)	$16.20	$40.88		

a. For income tax credit, use the annual average loss in capital value for the depreciation of line i, plus lines l, m, n, and r (labor, fuel, and lubrication costs). Depreciable portion of line i is normally the "cash boot" difference.

Fig. 8 — Comparison of Two Alternatives

Comparing Four Alternatives

A second example compares two additional options to the ones used in the first example (Fig. 9).

Option 3 is to purchase a combine that is 4 years old and own it for 10 more years.

Option 4 is to lease the same size combine for $140.00 per hour or $15,400 per year.

In Option 3, the depreciation formulas from Chapter 6, Table 1, are used to determine the initial cost and the "as-is" value when it has been owned 10 years. An estimate of the initial cost is made by first determining the "as-is" value at 4 years, then multiplying by 120% to arrive at the "as-is" value.

In Option 4, all of the expenses are tax deductible, and no finance charges are involved, assuming the rental cost would be paid at the time of harvest.

The four options represented by this example are the four most typical examples likely needed for making a decision. Option 4 could also be used for a custom operation. Enter the total annual cost in line q and the income tax deduction in line r under income tax credit, then compute cost per acre.

If the combine in Option 1 maintains its reliability and performance, the repair costs incurred with continued ownership can be justified. If repairs cannot restore reliability, resulting in unacceptable timeliness costs, then a different option should be considered.

Trading for a new combine means a large commitment of capital. If the ownership is extended to 12 years or more, cost per acre is likely to become more competitive. With the useful life of a combine set at 3,000 hours, longer ownership is the best way to lower cost per acre. At 110 hours a year, 20 years of ownership (2,200 hours) is still only a little over two-thirds of useful life.

Trading for a used machine will almost always result in a lower cost per acre than buying a new one. The problem may be one of finding a good used machine that will provide the needed reliability and performance. However, buying a used machine is a very good option if capital is short.

Renting may cost more per acre, but the capital outlay and risk is much less. In times of uncertainty, or when capital is limited, renting or using a custom operator is a desirable option.

Worksheet For Comparing Alternatives

Options →	1	2	3	4
Machine: Wheat Combine	Repair and Keep Using	Trade for New	Trade for Used	Rent for $140.00/Hour
a. Age at Purchase, Years	0	0	4	----
b. Current Age, Years	8	0	4	----
c. Ending Age, Years	18	10	14	----
d. Use, Acres/Year	800	800	800	800
e. Use, Hectares/Year	110	110	110	110
f. Original List Price	$100,000	$240,000	$200,000	$240,000
g. "As-Is" Value (Chapter 6, Table 1)	$40,841	$204,000	$124,000	----
h. "Cash Boot" Difference	----	$163,159	$83,159	----
i. Capital Value (Line g of Option 1 + Line h of Each Option)	$40,841	$204,000	$124,000	
j. Ending "As-Is" Value	$21,998	$85,560	$56,360	----
k. Loss in Capital Value	$18,843	$118,440	$67,640	----
l. Immediate Repairs	$7,000	$0	$0	----
m. Future Repairs	$13,670	$6,810	$10,980	----
n. Finance Charges on Investment Capital	$42,261	$159,258	$99,198	----
o. Taxes, Shelter, and Insurance Costs	$12,568	$57,910	$36,070	
p. Total (Lines k + l + m + n + o)	$94,432	$342,418	$213,888	----
q. Annual Costs, $/Year (Line p/[Line c – Line b])	$9,443	$34,242	$21,388	$15,400
r. Adjustments, $/Year				
Labor, $/Year	$1,100	$1,100	$1,100	$1,100
Fuel and Lubricant, $/Year	$3,520	$3,520	$3,520	$3,520
Age Penalty, $/Year	$1,500	$0	$750	$0
Income Tax Credit[a], $/Year	($2,607)	($6,161)	($4,255)	($4,189)
s. Total Cost, $/Year (Line q + r)	$12,956	$32,701	$22,503	$15,831
t. Total Cost, $/Acre (Line s/d)	$16.20	$40.88	$28.13	$19.79

a. For income tax credit, use the annual average loss in capital value for the depreciation of line i, plus lines l, m, n, and r (labor, fuel, and lubrication costs). Depreciable portion of line i is normally the "cash boot" difference.

Fig. 9 — Comparison of Four Alternatives

Summary

As stated at the beginning of the text, machinery management decisions are not easy and are made even more difficult by the lack of appropriate input data for localized applications. This text has used real-life examples for the primary purpose of teaching principles and methods. The greatest value in learning these methods and principles is to help you arrive at the best possible decisions for your own situations.

The procedure in this chapter will provide accurate answers for those decisions, provided you can determine the appropriate input data. The main advantage of this method is twofold:

- Fixed costs are based on real values.
- Credit can be given for income tax deductions.

Test Yourself

Questions

1. If you have an 8-year-old tractor that had a list price of $46,000 and your dealer offers to sell you a new one for a "cash boot" difference of $55,000, what is the capital value of the new tractor? Use "as-is" value for trade-in value for the 8-year-old tractor.

2. What would be the list price of the tractor in the previous problem if its value, or purchase price, is discounted 15% from list price?

3. You have just purchased a 4-year-old cotton picker that has 800 hours of use. If you use it 100 hours a year for the next 6 years, how much would you expect to spend for repairs? The original list price was $180,000.

4. If you finance $70,000 of the cost of the cotton picker for 5 years at 10% interest, what would be the total interest charge?

5. In the example problem (Fig. 9), what would be the average cost per acre if the combine in Option 2 had an ending "as-is" value of $70,000 instead of $50,522?

    a. $19.36

    b. $18.20

    c. $16.35

    d. $15.65

6. Option 1 of the example problem was "repair and keep." If the future repair costs were $6,000 instead of $13,670, what would be the total cost per acre?

    a. $9.86

    b. $10.11

    c. $10.65

    d. $11.92

7. Select a decision that is typical for your geographical area and use the form provided to compare two or more alternatives. One of these alternatives needs to be "repair and keep using." If necessary, visit a local farmer and obtain a good example to use. You may also need to visit with a local implement dealer to obtain an example.

Case Studies in Machinery Management

14

Introduction

Managing a farm or ranch business is a continuing process of learning, planning, evaluating, and decision-making. Each manager's goal must be to maximize returns to the business (Fig. 1).

Fig. 1 — Only Land Costs Are Higher Than Machinery Costs

Machinery management is an important part of farm and ranch business management. Only the cost for land exceeds the cost of owning and operating machinery. The case studies in this chapter will illustrate how decisions about machinery can affect the entire farm and ranch business program.

Chapter Objective

- Analyze and make decisions in case studies

Case Study — Record Keeping

Jack and Sue Johnson have farmed 800 acres in eastern Iowa for 10 years. They grow 400 acres of corn and 400 acres of soybeans each year. Now they have the opportunity to farm an additional 200 acres. Before taking on the additional acreage, they want to determine if they will be able to continue using the six-row equipment they already own. They own a 6-row planter and cultivator and a combine with a 6-row corn head and 20-foot grain platform.

From experience and their records, the Johnsons quickly determine that they could continue to use their present combine. Their records show they have been able to harvest both corn and soybeans in a timely manner, and they determine that an additional 200 acres would not present a problem with harvest. They have also determined that their present tractor is of sufficient size to pull a 12 row planter and also operate the hydraulics. They are concerned about whether the 6-row planter would be optimum for 1,200 acres.

In order to make a decision, the Johnsons evaluate three options:

Option 1 — Continue farming 800 acres with a six-row planter

Option 2 — Expand farm to 1,000 acres and use a six-row planter

Option 3 — Expand farm to 1,000 acres and purchase a new, larger planter

The Johnsons first estimate the fixed and variable costs of the planter for each option using their records and the methods presented in this book. For each option they obtain the list price and purchase price for the planter, and from their records they estimate the annual use and effective field capacity (Table 1).

Next they estimate the cost of the planter for each option assuming a length of ownership and using a cost factor from Table 5, Appendix. For a planter the cost factor includes fixed and repair costs. They estimate the planter cost to be $3.95 per acre for Option 1 and $3.61 per acre for Option 2 (Table 1). The lower cost for Option 2 is due to the increase in annual use of the planter.

For Option 3, the Johnsons consider a 12-row planter with a list price of $35,900. For the 12-row planter they estimate the effective field capacity to be 12.3 acres per hour, almost double the capacity of the 6-row planter. The 12-row planter would be used only 81 hours per year to plant 500 acres of corn and 500 acres of soybeans. With only 81 hours of use per year, they figure they could keep the planter for 10 years, which helps lower cost. The cost for the 12-row planter is estimated to be $5.10 per acre (Table 1).

The change in farm size or planter size affects the cost of the tractor used to pull the planter. The annual use of the tractor at present is 280 hours per year (Option 1). The Johnsons estimate the annual use of the tractor would increase to 352 hours per year for Option 2 and decrease to 198 hours per year for Option 3. Using these values for annual use and the other information for the tractor listed in Table 1, they obtain an estimated tractor cost for each option. Next, they add the tractor and planter costs and get estimated tractor plus planter costs of $8.82 per acre for Option 1, $7.81 per acre for Option 2, and $8.99 per acre for Option 3.

The only cost left is timeliness, a very important factor to be considered. To estimate timeliness costs, the Johnsons use their actual records of corn and soybean planting dates for Option 1. Their records show that they have consistently been able to get the crops planted without a timeliness penalty with their present farm size and 6-row planter. Therefore, the timeliness penalty is zero for Option 1. For Option 2, planting dates for both corn and soybeans would be late and they estimate a timeliness cost of $12.94 per acre. As expected for Option 3, with the 12-row planter, timeliness cost is zero.

For an estimate of total cost, the Johnsons add the tractor cost, planter cost, and timeliness cost for each option. The total cost estimates are $8.82 per acre for Option 1, $20.75 per acre for Option 2, and $8.11 per acre for Option 3. The Johnsons conclude that if they expand the farm by 200 acres, they would need a new planter because of the high cost of Option 2. Expanding the farm and buying a new 12-row planter would increase planting cost only $0.17 per acre ($8.99 − $8.82 = $0.17). They figure the $0.17 per acre extra cost would be offset by the increased crop production from 200 extra acres.

Analysis: This case study is an example of the benefits that can be obtained with a good record-keeping system. With good records, the Johnsons are able to accurately estimate the total cost for three options. With the cost estimates, they are able to make a sound decision on whether they can justify expanding their farm from 800 acres to 1,200 acres. Good records make it possible for them to base their decision on sound machinery management principles.

	Option 1	Option 2	Option 3
Farm Size, Acres =	800	1,000	1,000
Planter Size, Rows =	6	6	12
Planter			
List Price, $	17,290	17,290	35,900
Purchase Price, $	15,232	15,232	30,515
Annual Use, Hours	123	154	81
Effective Field Capacity, Acres/Hour	6.5	6.5	12.3
Length of Ownership, Years	7	5	10
Cost Factor, $/$1,000 of List Price[a]	1.432	1.3085	1.748
Fixed and Repair Costs, $/Hour	25.66	23.45	62.75
Fixed and Repair Costs, $/Acre	3.95	3.61	5.10
Tractor			
List Price, $	51,000	51,000	51,000
Purchase Price, $	43,350	43,350	43,350
Annual Use, Hours	280	352	198
Length of Ownership, Years	10	10	10
Cost Factor, $/$1,000 of List Price	0.445	0.360	0.605
Fixed and Repair Costs, $/Hour	22.70	18.36	30.86
Fixed and Repair Costs, $/Acre	3.49	2.82	2.51
Fuel Price, $/Gallon	4.00	4.00	4.00
Fuel Consumption, Gallon/Acre[b]	0.5	0.5	0.5
Fuel and Lubrication Costs, $/Acre	2.20	2.20	2.20
Total Tractor Costs, $/Acre	5.69	5.02	4.71
Tractor + Planter Cost			
Cost, $/Acre	9.64	8.63	9.81
Timeliness Costs			
Corn Planting Dates	4/25 to 5/15	4/25 to 5/22	4/25 to 5/8
Soybean Planting Dates	5/16 to 5/31	5/23 to 6/11	5/9 to 5/21
Timeliness Penalty Cost, $/Year	0	6,470	0
Timeliness Penalty Cost, $/Acre	0	12.94	0
Total Cost			
Total Cost, $/Acre	9.64	21.57	9.81

a. From Appendix, Table 5 for Planter and Table 1 for Tractor
b. From Chapter 7, Table 5

Table 1 — Determining Optimum Planter Size for the Johnsons

Case Study — Cost of New Machinery

Jim Renner has participated in a record-keeping service at his local lending institution for 10 years (Fig. 2). He has been very pleased with the service, particularly since it provides the kind of information needed by his tax consultant. His record-keeping service also provides a procedure for entering his machinery performance and cost data. For several years Jim didn't use this part of the system. He preferred to make machinery management decisions on his own. He liked to keep his equipment new and believed that frequent trading was necessary in order to take advantage of the investment tax credit.

Fig. 2 — Good Records Help You Make Management Decisions

His operation had been quite profitable in the earlier years, so he could use all of the investment credit. When the farm economy took a turn for the worse, he was faced with a declining net income and knew that he needed to cut costs. But where? He started keeping more complete records on machinery performance, costs of repairs and maintenance, and the cost to trade. The record-keeping service provided estimates on cost of ownership, based on a declining balance depreciation formula similar to the one in this text (Chapter 6).

After a few years of keeping more thorough records, Jim could see some ways to cut costs. The most serious problem that showed up from the records of the last few years was the difference between estimated and actual costs for the machinery. The record-keeping service assumed an ownership period of 10 years for his machinery; however, his own records indicated that he was trading in intervals of 3 to 5 years for most of the major machines. The differences in actual and projected costs for his combine and tractors are shown in Table 2.

	100 hp Tractor	150 hp Tractor	Combine
Annual Use, Hours per Year			
Estimated	600	500	200
Actual	550	450	150
Length of Ownership, Years			
Estimated	10	10	10
Actual	4	5	6
Cost New	$50,000	$72,000	$110,000
Fixed Costs Plus Repairs, per Year			
Estimated	$6,210	$8,942	$13,662
Actual	$8,772	$10,347	$15,428
Tax Credit, Prorated, Annual			
Estimated	$1,863	$2,683	$4,099
Actual	$2,632	$3,104	$4,628

Table 2 — Actual Cost vs. Estimated Cost Comparison

Actual costs exceed the estimated costs for the two tractors and the combine by a total of $5,733 a year before allowing for the investment tax credit. After allowing for the investment credit, the costs still exceed estimates by $4,014 per year. Annual use is actually less than estimated. As a result, Jim starts using the machinery record portion of his record-keeping service. He also decides not to trade as frequently as in the past. More effort will go into repairs and maintenance in order to maintain the reliability of his current machinery.

Analysis: A good record-keeping system is essential when decisions must be made. Good records also are necessary in order to make sure your income tax forms are correctly filled out. With his data, Jim could determine exactly what tax credits he could claim.

The money loss shown in this example would be much higher if investment tax credits were not allowed. If Jim had looked at these estimates carefully in the past and kept accurate records on his machinery costs, then he would have realized the effect of early trading on fixed costs. More importantly, Jim's situation shows the attitude often adopted by managers when income is good and there is less pressure to cut costs. If you are serious about being a top manager, it is important to use records to know and understand all you can about your farm business all of the time.

Fortunately, Jim understands the importance of complete records including those on his actual costs of owning and operating farm machinery. He knows that with the loss of the investment tax credit, longer ownership is going to be worth more than $5,000 a year. With longer ownership, it also will pay to have a good maintenance and repair program in order to maintain the reliability and avoid breakdowns of his current machinery.

Case Study — Financial Analysis

A financial analysis is a process used to determine the health of a farm or ranch business. The success of the farm manager often depends more on the use of a financial analysis than any other management tool. For this reason, financial statements must be considered as an integral part of farm and ranch management. A financial statement will do the following:

- Help obtain credit
- Determine net worth of the business and which way it is going
- Determine liquidity — the ability to convert assets to cash
- Determine solvency — measures the amount left after converting assets to cash and paying all debts
- Determine equity, particularly the ratio of debt to equity, another measure of solvency
- Determine borrowing power at any time

A financial analysis shows where you have been and where you are headed, financially. In other words, it can measure the success of the business.

Peer Pressure

Rick Young is 27 years old and has been lucky to get an early start in farming on his own. He took over from his father when he was 19 years old. He is an only child, has been married for 3 years to Jane, and loves to farm.

At first, he made the usual mistakes of a young, eager farmer. He worked hard, learned from his experiences, and soon gained a reputation for being a good farmer. He has a good mix of new and used equipment with plenty of capacity to get everything done on time. Within a few years he expanded his operation to 1,500 acres, mainly as a tenant. It seems that everyone wants him for a tenant. Besides 400 acres of his parents' land, he bought 90 acres with a farmstead, including an old house.

By using soil tests, quality seed, and proper chemicals, Rick's yields slowly increased. His family has always been conservative, and Rick continues that tradition. Instead of building a new house, he and Jane remodeled the old farmhouse on the land that they purchased. With the same attitude, Rick is always on the lookout for bargains in used equipment. He tries to keep his equipment inventory low in value but adequate to provide plenty of capacity in the field.

Rick does not belong to an outside record-keeping service, but he has a filing system where he can retrieve any information necessary for tax purposes or for making decisions. Increasing yields coupled with shrewd marketing and low machinery costs mean that he is in good shape financially. Other than loans for a storage bin and a combine, his only major debt is a mortgage on the 90 acres where the farmstead is located.

In the past, Rick has made most of his own decisions after consulting with family members. Today he is being influenced more and more by his neighbor Steve Sample. Steve is about his age and started farming about the same time. Steve is heavily in debt. He recently bought a new four-wheel drive tractor and a new house. Steve convinces Rick that he also should get a four-wheel drive tractor. He convinces Rick that it would save him time because of the added capacity. Rick likes the idea of more leisure time for him and Jane.

Rick first talks to his local dealer and finds that he can trade for the new tractor for $80,000 and his two older tractors. Rick next visits his lending institution. His lender, George Scott, is 60 years old. He has lived through the economic decline of the 1980s. George says, "Rick, I have known your family for a long time. I also know that you would qualify for the loan, but let's start with a financial statement just to see how you are doing." Rick works out the financial statement. He sees that his net worth has been steadily increasing. His cash flow is better than the average. Projections for the next 2 years show that he will be able to meet all operating costs and existing loan payments if prices remain at the level of the past 2 years.

The new loan of $80,000 for 6 years would mean an annual payment of $18,000. George says "While the $80,000 loan would still keep your equity-to-debt ratio within allowable limits, you would be left without much borrowing power for emergencies. If you were 30 years older and your business stabilized, then I would recommend that you ask for the loan. It would make life easier, but it would do very little, if anything, to increase your net income. You are still in a growth process and need to use your capital for productive assets. You also need a reserve for emergencies. If prices decline still further, you could be in serious trouble within 2 years."

The more George talks, the more Rick realizes what George is trying to tell him. He really doesn't need the big tractor. It would be used only 250 hours a year and the cost per acre would be higher than necessary. As a result, Rick trades the oldest tractor, which is 12 years old and needs extensive transmission work, for a new 150-horsepower tractor. This is about half the cost of the four-wheel drive tractor that he first thought he needed. Table 3 shows why the decision to buy the smaller tractor kept Rick and Jane in a much better financial position.

Available Borrowing Power	
Current	$90,000
After Trading for Four-Wheel Drive Tractor	$10,000
After Trading for Two-Wheel Drive Tractor	$50,000
Annual Loan Payments	
Four-Wheel Drive, 6 Year Loan	$18,000
Two-Wheel Drive, 6 Year Loan	$9,000

Table 3 — Comparing Available Borrowing Power and Annual Loan Payments

Analysis: Rick's situation is typical. He starts conservatively and stays out of heavy debt. Then as time goes on and he gains confidence and experience, he feels that he must keep up with his neighbors. George's advice is sound and reaffirms the need for him to deal with his own situation and not be influenced by his neighbors. By trading for the smaller tractor and continuing to work longer hours, he saves $9,000 a year. He also maintains a better reserve in borrowing power for emergencies. The bottom line is the fact that the four-wheel drive tractor would not increase income and may place Rick in a precarious financial position.

Case Study — Using a Budget

A budget is a plan for spending or saving money. Budgets take time if they are done properly, but they are invaluable to the farm manager. There are several kinds of budgets, with each having its place in the management of the farm business. A whole farm budget is a plan showing expenses and projected income for the entire farm, including all of the enterprises. An enterprise budget shows expenses and projected income for a single enterprise such as the production of cotton, wheat, cattle, or poultry. Sometimes a partial budget is used if only a portion of the projected income and expenses for an enterprise is needed (Fig. 3).

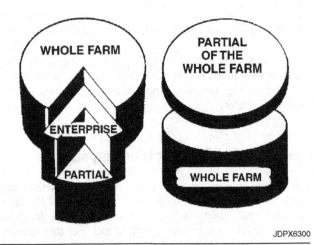

Fig. 3 — The Whole Farm Budget Can Be Divided Into Many Pieces

Budgets tell you if an enterprise is making a profit. Perhaps the greatest value in using budgets is to compare different plans or alternatives for an enterprise to determine which plan will show the greatest profit.

Dale Remington runs a large diversified farming operation in central Oklahoma. His operation includes cattle, hogs, wheat, cotton, and bermuda hay. His main source of income is from the cattle operation, which uses most of the bermuda hay produced on the farm.

He owns a conventional baler but uses custom hauling for his bermuda hay. He wants to continue bringing the bermuda hay program up to its productive potential, which will then allow for the expansion of the cattle operation.

Dale wants to look at an enterprise budget for using a loose hay stacking system instead of the conventional bale system. With the help of the local Extension Service he develops the enterprise budgets that follow, comparing the custom operation with owning his own loose hay stacking system.

The bermuda operation includes 100 acres that produce about 4.5 tons per acre.

Operating Inputs	Conventional Bales	Own a Loose Hay Stacking System
	\$ per Acre	
Fertilizer	38.60	38.60
Custom Haul	47.60	-------
Spray	2.60	2.60
Stand Establishment	8.00	8.00
Rent Fertilizer Spreader	1.25	1.25
Labor	10.35	12.98
Machinery Fuel, Lube, Repairs	22.49	32.21
Miscellaneous	17.14	10.95
Total Operating	148.03	106.59
Fixed Costs		
Machinery	54.01	62.95
Land	0.00	0.00
Total Fixed Cost	54.01	62.95
Income		
100 acres of bermuda, 4.5 tons/acre @ \$50.00/ton	225.00	225.00
Profit	22.96	55.46
Increased Profit, 100 acres x (\$55.46 − \$22.96) = \$3,250/year		

Table 4 — Enterprise Budgets

Analysis: This was a straightforward example of the use of budgets to help make a profitable management decision. The same approach could have been used to compare other alternatives, such as a big round bale system, along with the stacking wagon system. The key to success in using budgets is the accuracy of the inputs. It is much more difficult to accurately estimate machinery costs than other farm inputs such as feed, fertilizer, and chemicals. Again, the importance of using the information in this text, along with other farm management resources, is strongly encouraged.

Case Study — Cash Flow Analysis

There are two kinds of cash flow analysis. One is a cash flow summary, which is a financial history of business transactions over a given time period. The other is a cash flow projection, which is a future plan for money transactions over a specific period of time. Cash flow planning has become an increasingly valuable tool for managers. It is used to help increase capital at specific times of the year.

Increased capital may also mean that operating costs increase. Failure to have cash for paying operating costs is the first step to financial failure.

Cash flow projections can help the manager make decisions by choosing from a set of alternatives. Such is the situation with the following case study.

Mike White and his wife Kathy have been operating a 360 acre farm. They want to rent more land, including the 80 acres of good farmland only 2 miles down the road.

The additional land also means additional equipment, with the most critical item being the need for a larger tractor. Mike recently purchased a 100-hp tractor and must make payments of \$6,250 year for 4 more years.

After carefully reviewing their cash flow projections, they rule out the trade for a larger tractor. Their annual payments would go up by \$7,500 a year and eliminate all of their remaining borrowing power. This leaves them with three alternatives:

Option 1 — Buy a used tractor with approximately 100 hp. The cost of \$25,000 would require payments of \$6,000 a year.

Option 2 — Rent a 150-hp tractor for two months in the spring and one month in the fall. The cost would be \$1,250 per month for a total cost of \$3,750 per year.

Option 3 — Lease a new 100-hp tractor with the option to buy. The cost would bew \$7,500 per year for 5 years.

Analysis: Option 1 might be a good one if they can find a reliable used tractor that is not too old. Their cash flow also tells them that they cannot afford the risk of buying a used tractor and then having a major breakdown.

Option 3, the lease-purchase plan, has a 5-year commitment of \$7,500 a year, which is little better than owning.

By renting the tractor (Option 2), they have adequate borrowing power when they need it most. They have more cash to use for the purchase of needed tillage tools. By renting, they keep their options open, including the possibility of making a trade for a new 150-hp tractor in a few years.

Summary

Machinery management should not be separated from the total farm or ranch management system. The examples used in this chapter provide a brief look at the use of standard farm management procedures in making decisions involving farm machinery.

Learning to use these procedures will help you to make decisions and increase the chances for achieving production and profit goals for your farm or ranch. By combining proper management procedures with your own accurate record keeping, you will gain confidence in achieving financial and personal goals.

Test Yourself

Questions

1. What is the first step to determine the financial status of a farm business?
2. What would you do if a neighbor tries to convince you that you need a bigger tractor in order to get your work done faster?
3. Which of the financial procedures discussed in this chapter would be the most difficult to accomplish?
4. What can you learn from a cash flow projection?
5. Which procedure would you use to compare a minimum tillage program to conventional tillage?
6. After reviewing what you have learned from this chapter, develop a list of steps you can take to become a better farm manager.

Appendix

Fixed and Repair Cost Tables

All Wheel-Type Tractors

Useful life = 12,000 hours; Repair costs = $1,000 \times 0.006944 \times (hours/1,000)^2$; Remaining value; RV ($) = $1,000 \times 0.67 \times 0.94^y$; Taxes, shelter, insurance, and interest ($) = $0.13 \times RV$; Average accumulated cost per hour per $1,000 of list price (includes fixed costs and repairs, but does not include labor or fuel and lubricant costs). Machine is assumed to be purchased new at 85% of list price. Figures shown are average accumulated costs based on original ownership since machine was purchased.

The underlined portion of the table indicates that total machine usage exceeds the useful life. When useful life is reached, the annual repair cost is carried forward at the same rate for each of the following years. Values in this table do not include the effect of inflation over the period of ownership.

Hours of Annual Use	Cost per Hour per $1,000 of List Price — Age, Years (y)														
	1	2	3	4	5	6	7	8	9	10	11	12	13	14	15
200	1.655	1.129	0.942	0.841	0.775	0.726	0.688	0.656	0.629	0.605	0.584	0.565	0.547	0.531	0.516
250	1.325	0.904	0.756	0.676	0.623	0.585	0.555	0.530	0.509	0.490	0.474	0.459	0.446	0.434	0.422
300	1.104	0.755	0.632	0.565	0.522	0.491	0.467	0.447	0.430	0.415	0.402	0.390	0.380	0.370	0.362
350	0.947	0.648	0.543	0.487	0.451	0.425	0.404	0.388	0.374	0.362	0.352	0.342	0.334	0.326	0.320
400	0.830	0.569	0.477	0.429	0.398	0.376	0.358	0.345	0.333	0.323	0.315	0.307	0.301	0.295	0.289
450	0.738	0.507	0.426	0.384	0.357	0.338	0.323	0.312	0.302	0.294	0.287	0.281	0.276	0.271	0.267
500	0.665	0.457	0.386	0.348	0.325	0.308	0.296	0.286	0.278	0.271	0.266	0.261	0.257	0.253	0.250
550	0.605	0.417	0.353	0.319	0.298	0.284	0.273	0.265	0.258	0.253	0.249	0.245	0.242	0.240	0.237
600	0.555	0.384	0.325	0.295	0.277	0.264	0.255	0.248	0.243	0.239	0.235	0.233	0.231	0.229	0.228
650	0.513	0.355	0.302	0.275	0.259	0.248	0.240	0.235	0.230	0.227	0.225	0.223	0.222	0.221	0.220
700	0.477	0.331	0.283	0.258	0.244	0.234	0.228	0.223	0.220	0.218	0.216	0.215	0.214	0.214	0.214
750	0.446	0.311	0.266	0.244	0.231	0.223	0.217	0.214	0.211	0.210	0.209	0.209	0.209	0.209	0.210
800	0.419	0.293	0.251	0.231	0.220	0.213	0.208	0.206	0.204	0.203	0.203	0.204	0.205	0.206	0.207
850	0.395	0.277	0.238	0.220	0.210	0.204	0.201	0.199	0.198	0.198	0.199	0.200	0.201	0.203	0.205
900	0.374	0.263	0.227	0.211	0.202	0.197	0.194	0.193	0.193	0.194	0.195	0.197	0.199	0.201	0.203
950	0.355	0.250	0.217	0.202	0.195	0.191	0.189	0.189	0.189	0.190	0.192	0.195	0.197	0.199	0.201
1,000	0.338	0.239	0.208	0.195	0.188	0.185	0.184	0.185	0.186	0.188	0.190	0.193	0.195	0.197	0.198
1,100	0.308	0.220	0.193	0.183	0.178	0.176	0.177	0.178	0.181	0.184	0.187	0.190	0.192	0.193	0.194
1,200	0.284	0.204	0.181	0.173	0.170	0.170	0.171	0.174	0.178	0.182	0.185	0.187	0.189	0.190	0.191

Table 1 — All Wheel-Type Tractors

Crawler Tractors

Useful life = 16,000 hours; Repair costs = $1,000 \times 0.003125 \times (hours/1,000)^2$; Remaining value, RV ($) = $1,000 \times 0.67 \times 0.94^y$; Taxes, shelter, insurance, and interest ($) = $0.13 \times RV$; Average accumulated cost per hour per $1,000 of list price (includes fixed costs and repairs, but does not include labor or fuel and lubricant costs). Machine is assumed to be purchased new at 85% of list price. Figures shown are average accumulated costs based on original ownership since machine was purchased.

The underlined portion of the table indicates that total machine usage exceeds the useful life. When useful life is reached, the annual repair cost is carried forward at the same rate for each of the following years. Values in this table do not include the effect of inflation over the period of ownership.

Hours of Annual Use	Cost per Hour per $1,000 of List Price — Age, Years (y)														
	1	2	3	4	5	6	7	8	9	10	11	12	13	14	15
200	1.654	1.127	0.940	0.838	0.771	0.722	0.682	0.650	0.622	0.597	0.575	0.556	0.537	0.520	0.505
300	1.103	0.752	0.628	0.561	0.517	0.484	0.459	0.437	0.419	0.403	0.389	0.377	0.365	0.354	0.344
400	0.828	0.565	0.473	0.423	0.390	0.366	0.348	0.332	0.319	0.308	0.298	0.289	0.281	0.273	0.266
500	0.663	0.453	0.380	0.341	0.315	0.297	0.282	0.270	0.261	0.252	0.245	0.238	0.232	0.227	0.222
600	0.553	0.379	0.318	0.286	0.265	0.251	0.239	0.230	0.222	0.216	0.210	0.205	0.201	0.197	0.193
700	0.475	0.326	0.275	0.248	0.230	0.218	0.209	0.202	0.196	0.191	0.187	0.183	0.180	0.177	0.174
800	0.416	0.286	0.242	0.219	0.204	0.194	0.187	0.181	0.177	0.173	0.170	0.167	0.165	0.163	0.161
900	0.370	0.256	0.217	0.197	0.185	0.176	0.170	0.166	0.162	0.159	0.157	0.156	0.154	0.153	0.152
1,000	0.334	0.231	0.197	0.180	0.169	0.162	0.157	0.154	0.151	0.149	0.148	0.147	0.146	0.146	0.146
1,100	0.304	0.212	0.181	0.166	0.157	0.151	0.147	0.145	0.143	0.142	0.141	0.141	0.141	0.141	0.142
1,200	0.279	0.195	0.168	0.154	0.147	0.142	0.139	0.137	0.136	0.136	0.136	0.136	0.137	0.138	0.138
1,300	0.258	0.181	0.157	0.145	0.138	0.135	0.133	0.132	0.131	0.132	0.132	0.133	0.134	0.135	0.136
1,400	0.241	0.170	0.147	0.137	0.132	0.129	0.127	0.127	0.127	0.128	0.129	0.131	0.132	0.132	0.133
1,500	0.225	0.159	0.139	0.130	0.126	0.124	0.123	0.123	0.124	0.126	0.127	0.129	0.129	0.130	0.130
1,600	0.212	0.151	0.132	0.124	0.121	0.120	0.120	0.121	0.122	0.124	0.125	0.126	0.127	0.127	0.127
1,700	0.200	0.143	0.126	0.120	0.117	0.116	0.117	0.118	0.120	0.123	0.124	0.126	0.126	0.127	0.127
1,800	0.189	0.136	0.121	0.115	0.113	0.114	0.115	0.117	0.119	0.121	0.122	0.123	0.123	0.124	0.124
1,900	0.180	0.130	0.117	0.112	0.111	0.111	0.113	0.115	0.118	0.120	0.122	0.123	0.124	0.124	0.125
2,000	0.172	0.125	0.113	0.109	0.108	0.109	0.112	0.114	0.116	0.118	0.119	0.119	0.120	0.120	0.120

Table 2 — Crawler Tractors

Self-Propelled Combines Including Grain Heads

Useful life = 3,000 hours; Repair costs = $1,000 \times 0.03982 \times (hours/1,000)^2$; Remaining value, RV ($) = $1,000 \times 0.67 \times 0.94^y$; Taxes, shelter, insurance, and interest ($) = $0.13 \times RV$; Average accumulated cost per hour per $1,000 of list price (includes fixed costs and repairs, but does not include labor or fuel and lubricant costs). Machine is assumed to be purchased new at 85% of list price. Figures shown are average accumulated costs based on original ownership since machine was purchased.

The underlined portion of the table indicates that total machine usage exceeds the useful life. When useful life is reached, the annual repair cost is carried forward at the same rate for each of the following years. Values in this table do not include the effect of inflation over the period of ownership.

Cost per Hour per $1,000 of List Price															
Hours of Annual Use	Age, Years (y)														
	1	2	3	4	5	6	7	8	9	10	11	12	13	14	15
20	16.536	11.260	9.383	8.360	7.683	7.182	6.785	6.454	6.169	5.918	5.693	5.489	5.301	5.127	4.965
40	8.269	5.632	4.694	4.184	3.847	3.597	3.400	3.236	3.095	2.970	2.859	2.758	2.665	2.580	2.500
60	5.513	3.757	3.133	2.794	2.570	2.406	2.275	2.167	2.075	1.993	1.920	1.855	1.794	1.739	1.687
80	4.136	2.820	2.353	2.101	1.934	1.812	1.716	1.637	1.569	1.509	1.456	1.408	1.365	1.324	1.287
100	3.310	2.259	1.887	1.686	1.555	1.458	1.383	1.321	1.268	1.222	1.181	1.145	1.112	1.081	1.053
120	2.760	1.885	1.576	1.411	1.303	1.224	1.163	1.113	1.071	1.034	1.002	0.973	0.947	0.923	0.902
140	2.367	1.618	1.355	1.215	1.124	1.058	1.008	0.966	0.932	0.902	0.876	0.853	0.833	0.815	0.798
160	2.072	1.419	1.190	1.069	0.991	0.935	0.893	0.858	0.830	0.806	0.785	0.767	0.751	0.736	0.724
180	1.843	1.264	1.063	0.956	0.889	0.841	0.805	0.776	0.752	0.733	0.716	0.702	0.689	0.679	0.669
200	1.660	1.140	0.961	0.867	0.808	0.766	0.736	0.712	0.692	0.677	0.663	0.652	0.643	0.635	0.629
220	1.511	1.040	0.878	0.794	0.742	0.707	0.680	0.660	0.645	0.632	0.622	0.614	0.608	0.602	0.597
240	1.386	0.956	0.809	0.735	0.689	0.658	0.635	0.619	0.606	0.597	0.590	0.584	0.580	0.576	0.571
260	1.281	0.885	0.752	0.684	0.644	0.617	0.598	0.585	0.576	0.569	0.564	0.561	0.557	0.553	0.548
280	1.191	0.825	0.703	0.642	0.606	0.583	0.568	0.557	0.550	0.546	0.543	0.540	0.536	0.532	0.528
300	1.113	0.773	0.661	0.606	0.574	0.555	0.542	0.534	0.530	0.527	0.524	0.521	0.516	0.512	0.507
320	1.045	0.728	0.624	0.575	0.547	0.530	0.520	0.515	0.513	0.513	0.511	0.509	0.506	0.503	0.500
340	0.985	0.688	0.593	0.547	0.523	0.509	0.502	0.499	0.499	0.497	0.495	0.492	0.488	0.485	0.481
360	0.932	0.653	0.564	0.524	0.503	0.492	0.487	0.486	0.488	0.488	0.487	0.485	0.483	0.480	0.477
380	0.884	0.622	0.540	0.503	0.485	0.476	0.474	0.475	0.474	0.472	0.470	0.467	0.464	0.460	0.456
400	0.841	0.594	0.518	0.485	0.469	0.463	0.463	0.466	0.467	0.466	0.465	0.463	0.460	0.458	0.455

Table 3 — Self-Propelled Combines Including Grain Heads

Cotton Harvesters

Useful life = 5,000 hours; Repair costs = $1,000 \times 0.074044 \times (hours/1,000)^2$; Remaining value, RV ($) = $1,000 \times 0.67 \times 0.92^y$; Taxes, shelter, insurance, and interest ($) = $0.13 \times RV$; Average accumulated cost per hour per $1,000 of list price (includes fixed costs and repairs, but does not include labor or fuel and lubricant costs). Machine is assumed to be purchased new at 85% of list price. Figures shown are average accumulated costs based on original ownership since machine was purchased.

The underlined portion of the table indicates that total machine usage exceeds the useful life. When useful life is reached, the annual repair cost is carried forward at the same rate for each of the following years. Values in this table do not include the effect of inflation over the period of ownership.

Hours of Annual Use	Cost per Hour per $1,000 of List Price — Age, Years (y)														
	1	2	3	4	5	6	7	8	9	10	11	12	13	14	15
25	13.788	9.501	7.936	7.059	6.464	6.013	5.650	5.343	5.077	4.842	4.629	4.437	4.260	4.096	3.945
50	6.912	4.773	3.993	3.557	3.261	3.037	2.857	2.705	2.573	2.456	2.352	2.256	2.169	2.088	2.013
75	4.622	3.199	2.681	2.392	2.196	2.049	1.930	1.829	1.742	1.665	1.596	1.533	1.476	1.423	1.373
100	3.478	2.413	2.027	1.812	1.666	1.556	1.468	1.394	1.329	1.272	1.221	1.174	1.132	1.093	1.056
125	2.792	1.943	1.636	1.465	1.349	1.262	1.192	1.134	1.083	1.038	0.997	0.961	0.927	0.896	0.867
150	2.336	1.630	1.375	1.234	1.139	1.067	1.009	0.961	0.919	0.882	0.849	0.819	0.792	0.766	0.743
175	2.010	1.407	1.190	1.070	0.989	0.928	0.880	0.839	0.803	0.772	0.744	0.719	0.696	0.674	0.655
200	1.766	1.240	1.051	0.947	0.877	0.825	0.783	0.747	0.717	0.690	0.666	0.644	0.624	0.606	0.589
225	1.577	1.111	0.944	0.852	0.790	0.745	0.708	0.677	0.650	0.627	0.606	0.587	0.570	0.554	0.539
250	1.425	1.007	0.858	0.776	0.722	0.681	0.648	0.621	0.597	0.577	0.558	0.541	0.526	0.512	0.499
275	1.302	0.923	0.788	0.715	0.665	0.629	0.600	0.575	0.554	0.536	0.520	0.505	0.491	0.478	0.467
300	1.199	0.853	0.730	0.663	0.619	0.586	0.560	0.538	0.519	0.502	0.488	0.474	0.462	0.451	0.440
325	1.112	0.794	0.681	0.620	0.580	0.550	0.526	0.506	0.489	0.474	0.461	0.449	0.438	0.428	0.418
350	1.037	0.743	0.640	0.583	0.546	0.519	0.497	0.479	0.464	0.450	0.438	0.427	0.417	0.408	0.400
375	0.973	0.699	0.603	0.552	0.518	0.493	0.473	0.456	0.442	0.430	0.419	0.409	0.400	0.391	0.383
400	0.916	0.661	0.572	0.524	0.493	0.470	0.451	0.436	0.423	0.412	0.402	0.393	0.384	0.377	0.369
425	0.867	0.628	0.544	0.500	0.471	0.449	0.432	0.419	0.407	0.396	0.387	0.379	0.371	0.363	0.356
450	0.823	0.598	0.520	0.478	0.451	0.431	0.416	0.403	0.392	0.383	0.374	0.366	0.359	0.352	0.346
475	0.784	0.571	0.498	0.459	0.434	0.416	0.401	0.389	0.379	0.370	0.363	0.355	0.348	0.342	0.335
500	0.748	0.548	0.479	0.442	0.419	0.402	0.388	0.377	0.368	0.360	0.352	0.345	0.338	0.332	0.326

Table 4 — Cotton Harvesters

Planters and Drills

Useful life = 1,500 hours; Repair costs = $1,000 \times 0.32 \times (hours/1,000)^2$; Remaining value, RV ($) = $1,000 \times 0.67 \times 0.92^y$; Taxes, shelter, insurance, and interest ($) = $0.13 \times RV$; Average accumulated cost per hour per $1,000 of list price (includes fixed costs and repairs, but does not include labor or fuel and lubricant costs). Machine is assumed to be purchased new at 85% of list price. Figures shown are average accumulated costs based on original ownership since machine was purchased.

The underlined portion of the table indicates that total machine usage exceeds the useful life. When useful life is reached, the annual repair cost is carried forward at the same rate for each of the following years. Values in this table do not include the effect of inflation over the period of ownership.

Hours of Annual Use	Cost per Hour per $1,000 of List Price — Age, Years (y)														
	1	2	3	4	5	6	7	8	9	10	11	12	13	14	15
20	17.209	11.848	9.892	8.797	8.055	7.495	7.044	6.665	6.336	6.045	5.785	5.548	5.331	5.132	4.947
40	8.612	5.939	4.970	4.431	4.069	3.799	3.583	3.402	3.248	3.112	2.992	2.883	2.785	2.696	2.613
60	5.749	3.977	3.341	2.992	2.762	2.592	2.459	2.350	2.258	2.179	2.111	2.050	1.996	1.948	1.906
80	4.321	3.002	2.536	2.286	2.124	2.009	1.921	1.851	1.795	1.748	1.709	1.676	1.649	1.626	1.607
100	3.466	2.422	2.061	1.872	1.755	1.675	1.618	1.575	1.542	1.518	1.500	1.487	1.479	1.474	1.472
120	2.899	2.040	1.750	1.606	1.521	1.467	1.432	1.410	1.396	1.390	1.388	1.391	1.398	1.402	1.403
140	2.495	1.770	1.534	1.423	1.363	1.330	1.314	1.308	1.311	1.319	1.332	1.340	1.345	1.346	1.346
160	2.193	1.571	1.377	1.293	1.254	1.239	1.238	1.248	1.264	1.285	1.300	1.310	1.316	1.319	1.321
180	1.960	1.419	1.260	1.198	1.177	1.178	1.191	1.214	1.243	1.263	1.276	1.285	1.291	1.294	1.296
200	1.775	1.301	1.170	1.128	1.123	1.138	1.164	1.199	1.222	1.238	1.249	1.256	1.260	1.262	1.262
220	1.625	1.206	1.101	1.076	1.085	1.113	1.152	1.177	1.193	1.204	1.210	1.213	1.215	1.215	1.213
240	1.500	1.129	1.046	1.037	1.060	1.100	1.150	1.184	1.208	1.224	1.236	1.244	1.249	1.252	1.254
260	1.396	1.067	1.003	1.009	1.045	1.096	1.129	1.150	1.164	1.172	1.178	1.181	1.182	1.182	1.180
280	1.308	1.015	0.970	0.989	1.037	1.099	1.140	1.168	1.187	1.200	1.209	1.215	1.219	1.221	1.222
300	1.232	0.972	0.943	0.976	1.035	1.070	1.091	1.105	1.112	1.117	1.119	1.119	1.118	1.116	1.114
320	1.167	0.936	0.923	0.968	1.039	1.081	1.108	1.126	1.137	1.145	1.149	1.152	1.153	1.153	1.152
340	1.110	0.906	0.908	0.965	1.046	1.096	1.129	1.150	1.165	1.176	1.183	1.187	1.190	1.192	1.193
360	1.060	0.881	0.897	0.966	1.057	1.114	1.152	1.178	1.197	1.210	1.219	1.226	1.230	1.233	1.236
380	1.016	0.860	0.889	0.969	1.012	1.038	1.053	1.062	1.067	1.070	1.071	1.071	1.070	1.068	1.066
400	0.977	0.842	0.885	0.976	1.025	1.055	1.074	1.086	1.094	1.098	1.101	1.102	1.102	1.101	1.100

Table 5 — Planters and Drills

Moldboard Plows

Useful life = 2,000 hours; Repair costs = $1,000 \times 0.2873 \times (hours/1,000)^2$; Remaining value, RV ($) = $1,000 \times 0.67 \times 0.92^y$; Taxes, shelter, insurance, and interest ($) = $0.13 \times RV$; Average accumulated cost per hour per $1,000 of list price (includes fixed costs and repairs, but does not include labor or fuel and lubricant costs). Machine is assumed to be purchased new at 85% of list price. Figures shown are average accumulated costs based on original ownership since machine was purchased.

The underlined portion of the table indicates that total machine usage exceeds the useful life. When useful life is reached, the annual repair cost is carried forward at the same rate for each of the following years. Values in this table do not include the effect of inflation over the period of ownership.

Cost per Hour per $1,000 of List Price

Hours of Annual Use	Age, Years (y)														
	1	2	3	4	5	6	7	8	9	10	11	12	13	14	15
20	17.218	11.860	9.907	8.816	8.075	7.517	7.067	6.688	6.360	6.070	5.810	5.573	5.356	5.157	4.972
40	8.624	5.957	4.991	4.455	4.094	3.824	3.607	3.426	3.271	3.133	3.011	2.900	2.800	2.707	2.622
60	5.765	3.999	3.365	3.018	2.786	2.615	2.479	2.367	2.271	2.188	2.114	2.048	1.988	1.934	1.885
80	4.339	3.026	2.561	2.310	2.146	2.026	1.933	1.857	1.793	1.738	1.690	1.648	1.611	1.578	1.548
100	3.487	2.447	2.085	1.894	1.771	1.684	1.617	1.565	1.522	1.485	1.455	1.429	1.406	1.387	1.370
120	2.920	2.065	1.773	1.623	1.529	1.465	1.418	1.382	1.353	1.331	1.313	1.298	1.286	1.277	1.270
140	2.517	1.795	1.555	1.435	1.363	1.316	1.284	1.261	1.244	1.232	1.224	1.218	1.215	1.214	1.215
160	2.217	1.595	1.394	1.298	1.244	1.211	1.190	1.178	1.171	1.167	1.167	1.169	1.173	1.175	1.175
180	1.985	1.442	1.273	1.196	1.156	1.135	1.124	1.120	1.121	1.125	1.132	1.141	1.147	1.150	1.151
200	1.800	1.322	1.179	1.118	1.090	1.079	1.077	1.081	1.089	1.099	1.106	1.109	1.110	1.109	1.108
220	1.650	1.225	1.104	1.057	1.040	1.037	1.043	1.054	1.068	1.084	1.096	1.104	1.109	1.111	1.113
240	1.525	1.146	1.044	1.010	1.002	1.007	1.019	1.036	1.056	1.069	1.079	1.084	1.088	1.090	1.090
260	1.421	1.081	0.995	0.972	0.972	0.984	1.003	1.026	1.040	1.050	1.056	1.060	1.061	1.062	1.061
280	1.333	1.026	0.955	0.942	0.950	0.968	0.993	1.021	1.040	1.053	1.062	1.069	1.073	1.075	1.076
300	1.257	0.980	0.923	0.918	0.933	0.957	0.987	1.007	1.020	1.028	1.033	1.036	1.038	1.038	1.037
320	1.191	0.941	0.895	0.899	0.920	0.951	0.986	1.009	1.025	1.036	1.044	1.049	1.052	1.054	1.055
340	1.133	0.907	0.873	0.884	0.912	0.947	0.970	0.984	0.993	0.999	1.002	1.003	1.003	1.003	1.001
360	1.083	0.879	0.854	0.872	0.906	0.947	0.973	0.990	1.002	1.009	1.014	1.017	1.019	1.019	1.019
380	1.038	0.854	0.839	0.864	0.903	0.948	0.978	0.998	1.012	1.022	1.028	1.032	1.035	1.037	1.038
400	0.998	0.832	0.826	0.857	0.902	0.928	0.944	0.954	0.960	0.963	0.965	0.965	0.964	0.963	0.961

Table 6 — Moldboard Plows

Chisel Plows, Mulch Tillers, Disks, Cultivators, Harrows, etc.

Useful life = 2,000 hours; Repair costs = $1,000 \times 0.26524 \times (hours/1,000)^2$; Remaining value, RV ($) = $1,000 \times 0.67 \times 0.92^y$; Taxes, shelter, insurance, and interest ($) = $0.13 \times RV$; Average accumulated cost per hour per $1,000 of list price (includes fixed costs and repairs, but does not include labor or fuel and lubricant costs). Machine is assumed to be purchased new at 85% of list price. Figures shown are average accumulated costs based on original ownership since machine was purchased.

The underlined portion of the table indicates that total machine usage exceeds the useful life. When useful life is reached, the annual repair cost is carried forward at the same rate for each of the following years. Values in this table do not include the effect of inflation over the period of ownership.

	Cost per Hour per $1,000 of List Price														
Hours of Annual Use	Age, Years (y)														
	1	2	3	4	5	6	7	8	9	10	11	12	13	14	15
20	17.260	11.912	9.963	8.874	8.135	7.578	7.128	6.749	6.421	6.130	5.869	5.631	5.413	5.212	5.026
40	8.676	6.016	5.052	4.516	4.154	3.882	3.663	3.479	3.320	3.179	3.053	2.938	2.834	2.737	2.647
60	5.821	4.060	3.426	3.076	2.841	2.664	2.523	2.405	2.303	2.213	2.133	2.060	1.993	1.932	1.875
80	4.398	3.087	2.619	2.363	2.191	2.064	1.962	1.877	1.804	1.740	1.683	1.631	1.584	1.541	1.501
100	3.547	2.507	2.139	1.939	1.807	1.709	1.631	1.567	1.512	1.463	1.420	1.382	1.346	1.314	1.284
120	2.981	2.123	1.822	1.661	1.555	1.477	1.415	1.365	1.321	1.284	1.250	1.220	1.193	1.169	1.146
140	2.579	1.851	1.599	1.464	1.377	1.314	1.264	1.224	1.189	1.159	1.133	1.109	1.088	1.069	1.051
160	2.278	1.648	1.432	1.319	1.246	1.194	1.153	1.121	1.093	1.069	1.048	1.029	1.013	0.997	0.981
180	2.045	1.492	1.305	1.208	1.147	1.103	1.070	1.043	1.020	1.001	0.985	0.970	0.956	0.942	0.928
200	1.860	1.368	1.204	1.120	1.068	1.032	1.004	0.982	0.964	0.949	0.934	0.920	0.906	0.892	0.880
220	1.709	1.267	1.123	1.050	1.006	0.975	0.952	0.935	0.920	0.908	0.896	0.884	0.873	0.862	0.851
240	1.584	1.184	1.056	0.992	0.954	0.929	0.910	0.896	0.885	0.873	0.862	0.851	0.840	0.829	0.819
260	1.478	1.115	1.000	0.945	0.912	0.891	0.876	0.865	0.854	0.842	0.831	0.820	0.810	0.800	0.790
280	1.388	1.056	0.953	0.904	0.877	0.860	0.848	0.839	0.830	0.821	0.811	0.802	0.793	0.784	0.776
300	1.311	1.005	0.913	0.870	0.847	0.833	0.824	0.814	0.804	0.795	0.785	0.776	0.767	0.758	0.749
320	1.243	0.962	0.878	0.841	0.822	0.811	0.804	0.797	0.788	0.780	0.772	0.764	0.756	0.748	0.740
340	1.184	0.924	0.848	0.816	0.800	0.792	0.783	0.773	0.764	0.755	0.746	0.737	0.729	0.721	0.713
360	1.132	0.890	0.822	0.795	0.782	0.776	0.768	0.761	0.753	0.745	0.737	0.729	0.721	0.714	0.707
380	1.086	0.861	0.799	0.776	0.766	0.762	0.756	0.750	0.743	0.736	0.729	0.722	0.715	0.709	0.703
400	1.044	0.835	0.779	0.759	0.751	0.743	0.735	0.726	0.717	0.709	0.701	0.694	0.686	0.679	0.673

Table 7 — Chisel Plows, Mulch Tillers, Disks, Cultivators, Harrows, etc.

Mowers

Useful life = 2,000 hours; Repair costs = $1,000 \times 0.4617 \times (hours/1,000)^2$; Remaining value. RV ($) = $1,000 \times 0.67 \times 0.92^y$; Taxes, shelter, insurance, and interest ($) = $0.13 \times RV$; Average accumulated cost per hour per $1,000 of list price (includes fixed costs and repairs, but does not include labor or fuel and lubricant costs). Machine is assumed to be purchased new at 85% of list price. Figures shown are average accumulated costs based on original ownership since machine was purchased.

The underlined portion of the table indicates that total machine usage exceeds the useful life. When useful life is reached, the annual repair cost is carried forward at the same rate for each of the following years. Values in this table do not include the effect of inflation over the period of ownership.

Cost per Hour per $1,000 of List Price

Hours of Annual Use	1	2	3	4	5	6	7	8	9	10	11	12	13	14	15
20	17.235	11.887	9.942	8.856	8.122	7.569	7.124	6.750	6.426	6.140	5.884	5.651	5.438	5.242	5.061
40	8.651	5.998	5.043	4.517	4.165	3.902	3.693	3.519	3.370	3.239	3.122	3.017	2.921	2.834	2.754
60	5.799	4.051	3.431	3.096	2.875	2.714	2.587	2.484	2.396	2.320	2.253	2.194	2.141	2.093	2.050
80	4.380	3.088	2.639	2.402	2.251	2.142	2.060	1.993	1.939	1.893	1.853	1.819	1.789	1.763	1.740
100	3.533	2.517	2.174	1.999	1.890	1.816	1.761	1.719	1.686	1.660	1.638	1.621	1.607	1.595	1.586
120	2.972	2.143	1.872	1.739	1.661	1.611	1.577	1.552	1.535	1.523	1.515	1.510	1.507	1.506	1.507
140	2.574	1.881	1.663	1.562	1.507	1.475	1.456	1.446	1.441	1.440	1.442	1.447	1.453	1.461	1.471
160	2.279	1.688	1.511	1.435	1.399	1.382	1.376	1.377	1.382	1.390	1.401	1.414	1.428	1.439	1.446
180	2.051	1.541	1.397	1.342	1.321	1.317	1.321	1.332	1.346	1.362	1.381	1.401	1.415	1.426	1.435
200	1.870	1.427	1.311	1.273	1.265	1.271	1.285	1.304	1.325	1.349	1.366	1.378	1.387	1.392	1.396
220	1.724	1.336	1.243	1.220	1.224	1.239	1.262	1.288	1.316	1.346	1.369	1.386	1.398	1.408	1.415
240	1.604	1.263	1.190	1.180	1.194	1.218	1.248	1.281	1.316	1.341	1.360	1.374	1.384	1.392	1.397
260	1.503	1.203	1.148	1.150	1.172	1.204	1.241	1.280	1.308	1.328	1.343	1.353	1.361	1.366	1.370
280	1.418	1.153	1.114	1.127	1.158	1.197	1.240	1.285	1.317	1.341	1.359	1.373	1.383	1.390	1.396
300	1.346	1.112	1.087	1.110	1.149	1.194	1.243	1.277	1.301	1.318	1.331	1.340	1.347	1.351	1.354
320	1.283	1.078	1.066	1.097	1.143	1.195	1.250	1.288	1.316	1.336	1.351	1.362	1.371	1.377	1.382
340	1.229	1.049	1.049	1.089	1.142	1.200	1.238	1.264	1.282	1.295	1.304	1.311	1.315	1.318	1.320
360	1.182	1.025	1.036	1.084	1.143	1.206	1.249	1.278	1.299	1.314	1.325	1.333	1.339	1.343	1.346
380	1.140	1.004	1.026	1.081	1.146	1.215	1.261	1.294	1.317	1.334	1.347	1.357	1.364	1.369	1.373
400	1.103	0.987	1.018	1.080	1.152	1.196	1.224	1.244	1.257	1.267	1.273	1.277	1.280	1.282	1.282

Table 8 — Mowers

Large and Small Square Balers

Useful life = 2,500 hours; Repair costs = $1,000 \times 0.1441 \times (hours/1,000)^2$; Remaining value, RV ($) = $1,000 \times 0.67 \times 0.92^y$; Taxes, shelter, insurance, and interest ($) = $0.13 \times RV$; Average accumulated cost per hour per $1,000 of list price (includes fixed costs and repairs, but does not include labor or fuel and lubricant costs). Machine is assumed to be purchased new at 85% of list price. Figures shown are average accumulated costs based on original ownership since machine was purchased.

The underlined portion of the table indicates that total machine usage exceeds the useful life. When useful life is reached, the annual repair cost is carried forward at the same rate for each of the following years. Values in this table do not include the effect of inflation over the period of ownership.

	Cost per Hour per $1,000 of List Price														
Hours of Annual Use	Age, Years (y)														
	1	2	3	4	5	6	7	8	9	10	11	12	13	14	15
20	17.211	11.850	9.892	8.797	8.053	7.491	7.037	6.655	6.324	6.031	5.767	5.527	5.308	5.105	4.917
40	8.613	5.938	4.965	4.422	4.055	3.778	3.556	3.369	3.207	3.065	2.937	2.821	2.715	2.617	2.527
60	5.750	3.973	3.329	2.972	2.732	2.552	2.408	2.287	2.184	2.093	2.011	1.938	1.871	1.810	1.753
80	4.320	2.993	2.515	2.252	2.077	1.946	1.842	1.756	1.683	1.618	1.561	1.510	1.463	1.421	1.382
100	3.464	2.407	2.030	1.825	1.689	1.589	1.510	1.445	1.390	1.342	1.300	1.263	1.229	1.199	1.172
120	2.894	2.019	1.710	1.543	1.434	1.355	1.293	1.243	1.201	1.165	1.134	1.106	1.082	1.060	1.041
140	2.488	1.743	1.483	1.345	1.255	1.192	1.143	1.104	1.072	1.044	1.021	1.001	0.984	0.969	0.955
160	2.184	1.538	1.315	1.198	1.124	1.073	1.034	1.003	0.979	0.959	0.942	0.928	0.916	0.906	0.898
180	1.948	1.379	1.185	1.086	1.025	0.983	0.952	0.929	0.911	0.896	0.885	0.876	0.869	0.863	0.857
200	1.760	1.253	1.083	0.998	0.947	0.913	0.889	0.872	0.859	0.850	0.843	0.838	0.835	0.831	0.826
220	1.607	1.151	1.001	0.928	0.886	0.859	0.841	0.829	0.820	0.815	0.813	0.812	0.809	0.806	0.802
240	1.480	1.067	0.934	0.871	0.836	0.815	0.802	0.795	0.791	0.790	0.790	0.789	0.787	0.784	0.780
260	1.373	0.996	0.878	0.824	0.795	0.780	0.772	0.768	0.768	0.770	0.770	0.769	0.766	0.763	0.759
280	1.281	0.936	0.831	0.785	0.762	0.751	0.747	0.748	0.751	0.751	0.750	0.748	0.744	0.740	0.736
300	1.202	0.885	0.791	0.752	0.735	0.728	0.728	0.732	0.738	0.741	0.742	0.742	0.740	0.738	0.735
320	1.133	0.841	0.757	0.724	0.712	0.709	0.713	0.720	0.723	0.723	0.723	0.721	0.718	0.714	0.711
340	1.073	0.802	0.727	0.701	0.693	0.694	0.701	0.710	0.716	0.719	0.720	0.719	0.718	0.716	0.713
360	1.019	0.768	0.702	0.681	0.677	0.682	0.691	0.696	0.698	0.698	0.696	0.694	0.691	0.688	0.684
380	0.972	0.739	0.680	0.663	0.663	0.671	0.684	0.691	0.695	0.696	0.696	0.695	0.693	0.691	0.688
400	0.929	0.712	0.661	0.649	0.652	0.664	0.679	0.688	0.694	0.697	0.698	0.698	0.697	0.695	0.693

Table 9 — Large and Small Square Balers

Large Round Balers

Useful life = 1,500 hours; Repair costs = $1,000 \times 0.434 \times (hours/1,000)^2$; Remaining value, RV ($) = $1,000 \times 0.67 \times 0.92^y$; Taxes, shelter, insurance, and interest ($) = 0.13 × RV; Average accumulated cost per hour per $1,000 of list price (includes fixed costs and repairs, but does not include labor or fuel and lubricant costs). Machine is assumed to be purchased new at 85% of list price. Figures shown are average accumulated costs based on original ownership since machine was purchased.

The underlined portion of the table indicates that total machine usage exceeds the useful life. When useful life is reached, the annual repair cost is carried forward at the same rate for each of the following years. Values in this table do not include the effect of inflation over the period of ownership.

Hours of Annual Use	Cost per Hour per $1,000 of List Price — Age, Years (y)														
	1	2	3	4	5	6	7	8	9	10	11	12	13	14	15
20	17.224	11.872	9.923	8.835	8.099	7.544	7.097	6.722	6.398	6.111	5.853	5.620	5.406	5.210	5.028
40	8.636	5.977	5.018	4.489	4.135	3.871	3.660	3.485	3.335	3.204	3.087	2.982	2.887	2.799	2.719
60	5.781	4.026	3.402	3.064	2.842	2.680	2.553	2.449	2.361	2.285	2.219	2.161	2.109	2.062	2.020
80	4.359	3.060	2.608	2.369	2.216	2.107	2.025	1.959	1.906	1.861	1.823	1.790	1.762	1.738	1.718
100	3.510	2.487	2.141	1.964	1.855	1.781	1.728	1.687	1.656	1.632	1.613	1.598	1.587	1.579	1.573
120	2.947	2.112	1.838	1.704	1.627	1.578	1.545	1.524	1.509	1.501	1.496	1.495	1.496	1.495	1.491
140	2.548	1.848	1.628	1.527	1.473	1.444	1.428	1.421	1.420	1.424	1.431	1.434	1.434	1.431	1.428
160	2.251	1.654	1.476	1.401	1.367	1.353	1.351	1.357	1.367	1.381	1.389	1.394	1.396	1.395	1.393
180	2.022	1.507	1.363	1.309	1.291	1.291	1.301	1.317	1.337	1.350	1.358	1.362	1.364	1.364	1.362
200	1.840	1.392	1.276	1.241	1.237	1.249	1.269	1.294	1.311	1.321	1.327	1.330	1.330	1.330	1.328
220	1.693	1.301	1.209	1.190	1.198	1.221	1.250	1.268	1.279	1.286	1.289	1.289	1.288	1.286	1.283
240	1.572	1.228	1.157	1.152	1.171	1.203	1.241	1.266	1.283	1.294	1.301	1.305	1.307	1.307	1.306
260	1.471	1.168	1.116	1.123	1.153	1.194	1.219	1.234	1.243	1.248	1.251	1.251	1.250	1.248	1.246
280	1.386	1.119	1.083	1.102	1.142	1.190	1.222	1.242	1.255	1.264	1.269	1.272	1.273	1.273	1.272
300	1.313	1.078	1.057	1.087	1.136	1.163	1.180	1.189	1.194	1.196	1.196	1.195	1.192	1.190	1.186
320	1.250	1.044	1.037	1.077	1.134	1.168	1.188	1.201	1.209	1.213	1.215	1.216	1.215	1.214	1.212
340	1.195	1.015	1.022	1.071	1.136	1.175	1.200	1.216	1.226	1.233	1.237	1.239	1.240	1.240	1.239
360	1.147	0.991	1.010	1.069	1.141	1.185	1.214	1.233	1.246	1.254	1.260	1.264	1.266	1.267	1.267
380	1.106	0.972	1.002	1.069	1.104	1.124	1.136	1.142	1.145	1.146	1.146	1.145	1.142	1.140	1.137
400	1.069	0.955	0.996	1.071	1.112	1.135	1.150	1.158	1.163	1.166	1.166	1.166	1.165	1.163	1.161

Table 10 — Large Round Balers

Self-Propelled Forage Harvesters

Useful life = 4,000 hours; Repair costs = $1,000 \times 0.03125 \times (hours/1,000)^2$; Remaining value, RV ($) = $1,000 \times 0.67 \times 0.9?^y$; Taxes, shelter, insurance, and interest ($) = $0.13 \times RV$; Average accumulated cost per hour per $1,000 of list price (includes fixed costs and repairs, but does not include labor or fuel and lubricant costs). Machine is assumed to be purchased new at 85% of list price. Figures shown are average accumulated costs based on original ownership since machine was purchased.

The underlined portion of the table indicates that total machine usage exceeds the useful life. When useful life is reached, the annual repair cost is carried forward at the same rate for each of the following years. Values in this table do not include the effect of inflation over the period of ownership.

Hours of Annual Use	\multicolumn{15}{c}{Cost per Hour per $1,000 of List Price — Age, Years (y)}														
	1	2	3	4	5	6	7	8	9	10	11	12	13	14	15
20	17.876	12.406	10.352	9.169	8.348	7.716	7.200	6.762	6.380	6.042	5.738	5.462	5.211	4.980	4.766
40	8.939	6.205	5.179	4.588	4.179	3.864	3.606	3.388	3.198	3.030	2.879	2.742	2.618	2.503	2.397
60	5.960	4.139	3.456	3.063	2.791	2.582	2.412	2.267	2.142	2.031	1.931	1.841	1.759	1.683	1.614
80	4.471	3.106	2.595	2.302	2.099	1.943	1.816	1.709	1.616	1.534	1.460	1.394	1.333	1.278	1.227
100	3.578	2.487	2.079	1.846	1.685	1.561	1.461	1.376	1.303	1.238	1.181	1.128	1.081	1.038	0.998
120	2.983	2.075	1.736	1.543	1.410	1.308	1.225	1.156	1.096	1.043	0.996	0.954	0.916	0.881	0.849
140	2.558	1.781	1.492	1.327	1.214	1.128	1.059	1.000	0.950	0.906	0.867	0.832	0.800	0.771	0.745
160	2.239	1.561	1.309	1.166	1.068	0.994	0.934	0.885	0.842	0.804	0.771	0.742	0.715	0.691	0.670
180	1.992	1.390	1.167	1.041	0.955	0.891	0.839	0.796	0.759	0.727	0.699	0.674	0.651	0.631	0.613
200	1.794	1.253	1.054	0.942	0.866	0.809	0.763	0.726	0.694	0.666	0.642	0.620	0.602	0.585	0.569
220	1.632	1.141	0.962	0.861	0.793	0.742	0.702	0.669	0.641	0.617	0.597	0.578	0.562	0.548	0.536
240	1.497	1.049	0.885	0.794	0.733	0.688	0.652	0.623	0.599	0.578	0.560	0.545	0.531	0.519	0.509
260	1.383	0.970	0.821	0.738	0.683	0.642	0.610	0.585	0.563	0.546	0.530	0.517	0.506	0.496	0.488
280	1.286	0.904	0.766	0.690	0.640	0.603	0.575	0.553	0.534	0.519	0.506	0.495	0.485	0.478	0.471
300	1.201	0.846	0.718	0.649	0.603	0.570	0.545	0.525	0.509	0.496	0.485	0.476	0.469	0.463	0.457
320	1.127	0.795	0.677	0.613	0.572	0.542	0.520	0.502	0.488	0.477	0.468	0.461	0.455	0.449	0.443
340	1.062	0.751	0.641	0.582	0.544	0.517	0.498	0.482	0.471	0.461	0.454	0.448	0.443	0.437	0.431
360	1.004	0.712	0.609	0.554	0.520	0.496	0.478	0.465	0.455	0.448	0.442	0.438	0.434	0.429	0.424
380	0.953	0.677	0.580	0.530	0.499	0.477	0.462	0.451	0.442	0.436	0.432	0.428	0.423	0.418	0.413
400	0.906	0.645	0.555	0.508	0.480	0.461	0.447	0.438	0.431	0.427	0.422	0.416	0.411	0.406	0.400

Table 11 — Self-Propelled Forage Harvesters

Self-Propelled Windrowers

Useful life = 3,000 hours; Repair costs = $1,000 \times 0.0611 \times (hours/1,000)^2$; Remaining value, RV ($) = $1,000 \times 0.67 \times 0.92^y$; Taxes, shelter, insurance, and interest ($) = 0.13 × RV; Average accumulated cost per hour per $1,000 of list price (includes fixed costs and repairs, but does not include labor or fuel and lubricant costs). Machine is assumed to be purchased new at 85% of list price. Figures shown are average accumulated costs based on original ownership since machine was purchased.

The underlined portion of the table indicates that total machine usage exceeds the useful life. When useful life is reached, the annual repair cost is carried forward at the same rate for each of the following years. Values in this table do not include the effect of inflation over the period of ownership.

Cost per Hour per $1,000 of List Price

Hours of Annual Use	Age, Years (y)														
	1	2	3	4	5	6	7	8	9	10	11	12	13	14	15
20	17.206	11.841	9.881	8.782	8.036	7.472	7.016	6.632	6.298	6.003	5.737	5.496	5.275	5.070	4.880
40	8.605	5.924	4.946	4.398	4.027	3.747	3.521	3.331	3.166	3.020	2.889	2.770	2.661	2.561	2.468
60	5.739	3.954	3.303	2.940	2.695	2.510	2.361	2.237	2.129	2.034	1.948	1.871	1.801	1.736	1.676
80	4.306	2.969	2.484	2.214	2.032	1.895	1.786	1.695	1.616	1.547	1.485	1.429	1.378	1.332	1.289
100	3.447	2.380	1.994	1.780	1.637	1.530	1.444	1.373	1.312	1.259	1.212	1.170	1.131	1.096	1.064
120	2.875	1.988	1.668	1.492	1.375	1.288	1.219	1.162	1.114	1.072	1.035	1.002	0.972	0.945	0.920
140	2.466	1.708	1.437	1.288	1.190	1.118	1.061	1.014	0.975	0.941	0.912	0.886	0.862	0.842	0.823
160	2.160	1.499	1.264	1.136	1.053	0.992	0.944	0.906	0.874	0.847	0.823	0.802	0.784	0.768	0.754
180	1.923	1.337	1.130	1.019	0.947	0.895	0.856	0.824	0.798	0.776	0.757	0.741	0.727	0.715	0.705
200	1.733	1.208	1.024	0.927	0.864	0.820	0.786	0.760	0.739	0.721	0.707	0.695	0.685	0.676	0.670
220	1.578	1.103	0.938	0.852	0.797	0.759	0.731	0.710	0.693	0.679	0.668	0.660	0.653	0.648	0.642
240	1.448	1.016	0.867	0.790	0.742	0.710	0.687	0.669	0.656	0.646	0.638	0.633	0.629	0.624	0.619
260	1.339	0.942	0.807	0.739	0.697	0.669	0.650	0.636	0.627	0.620	0.615	0.612	0.609	0.604	0.600
280	1.246	0.880	0.757	0.695	0.659	0.636	0.620	0.610	0.603	0.599	0.597	0.594	0.590	0.586	0.581
300	1.165	0.826	0.713	0.658	0.627	0.608	0.595	0.588	0.584	0.583	0.580	0.576	0.572	0.567	0.562
320	1.095	0.779	0.676	0.627	0.600	0.584	0.575	0.570	0.569	0.570	0.569	0.567	0.565	0.562	0.558
340	1.033	0.738	0.643	0.599	0.576	0.564	0.558	0.556	0.557	0.556	0.554	0.551	0.547	0.544	0.539
360	0.978	0.702	0.615	0.576	0.556	0.547	0.543	0.544	0.547	0.548	0.548	0.546	0.544	0.542	0.538
380	0.929	0.670	0.590	0.555	0.539	0.532	0.531	0.534	0.535	0.534	0.531	0.528	0.525	0.521	0.517
400	0.885	0.641	0.567	0.537	0.524	0.520	0.521	0.527	0.529	0.529	0.528	0.527	0.524	0.521	0.518

Table 12 — Self-Propelled Windrowers

Rakes

Useful life = 2,500 hours; Repair costs = $1,000 \times 0.1663 \times (hours/1,000)^2$; Remaining value, RV ($) = $1,000 \times 0.67 \times 0.92^y$; Taxes, shelter, insurance, and interest ($) = $0.13 \times RV$; Average accumulated cost per hour per $1,000 of list price (includes fixed costs and repairs, but does not include labor or fuel and lubricant costs). Machine is assumed to be purchased new at 85% of list price. Figures shown are average accumulated costs based on original ownership since machine was purchased.

The underlined portion of the table indicates that total machine usage exceeds the useful life. When useful life is reached, the annual repair cost is carried forward at the same rate for each of the following years. Values in this table do not include the effect of inflation over the period of ownership.

Cost per Hour per $1,000 of List Price

Hours of Annual Use	1	2	3	4	5	6	7	8	9	10	11	12	13	14	15
20	17.240	11.884	9.931	8.838	8.096	7.536	7.083	6.702	6.371	6.078	5.815	5.575	5.356	5.153	4.965
40	8.648	5.980	5.010	4.469	4.102	3.826	3.604	3.416	3.254	3.111	2.982	2.865	2.757	2.658	2.567
60	5.789	4.017	3.376	3.020	2.779	2.599	2.453	2.331	2.226	2.133	2.049	1.973	1.903	1.839	1.780
80	4.362	3.040	2.563	2.300	2.123	1.990	1.884	1.795	1.718	1.650	1.589	1.534	1.484	1.437	1.394
100	3.507	2.455	2.078	1.871	1.732	1.628	1.546	1.477	1.417	1.364	1.318	1.275	1.236	1.201	1.168
120	2.939	2.067	1.757	1.587	1.474	1.390	1.323	1.267	1.219	1.177	1.140	1.106	1.075	1.047	1.021
140	2.534	1.791	1.529	1.386	1.291	1.221	1.166	1.120	1.081	1.046	1.015	0.988	0.963	0.940	0.918
160	2.231	1.585	1.359	1.236	1.156	1.097	1.050	1.011	0.978	0.950	0.924	0.901	0.880	0.861	0.844
180	1.995	1.426	1.227	1.121	1.052	1.001	0.961	0.928	0.900	0.876	0.855	0.835	0.818	0.802	0.787
200	1.808	1.299	1.123	1.030	0.969	0.925	0.891	0.863	0.839	0.819	0.800	0.784	0.770	0.756	0.742
220	1.655	1.196	1.039	0.956	0.903	0.864	0.835	0.810	0.790	0.773	0.757	0.744	0.730	0.718	0.706
240	1.528	1.111	0.969	0.895	0.848	0.814	0.789	0.768	0.750	0.735	0.722	0.710	0.698	0.686	0.675
260	1.420	1.039	0.910	0.844	0.802	0.773	0.750	0.732	0.717	0.705	0.692	0.680	0.669	0.658	0.648
280	1.329	0.977	0.861	0.801	0.764	0.738	0.718	0.703	0.690	0.677	0.666	0.654	0.644	0.633	0.623
300	1.250	0.925	0.818	0.764	0.731	0.708	0.691	0.678	0.667	0.656	0.646	0.636	0.626	0.617	0.608
320	1.181	0.879	0.781	0.732	0.703	0.682	0.668	0.656	0.645	0.634	0.624	0.614	0.605	0.596	0.587
340	1.120	0.839	0.749	0.704	0.678	0.660	0.647	0.638	0.628	0.619	0.610	0.601	0.592	0.584	0.576
360	1.066	0.804	0.720	0.680	0.656	0.641	0.630	0.619	0.609	0.599	0.590	0.581	0.573	0.565	0.557
380	1.018	0.772	0.695	0.659	0.638	0.624	0.615	0.606	0.596	0.588	0.579	0.571	0.563	0.556	0.549
400	0.976	0.744	0.673	0.640	0.621	0.609	0.601	0.593	0.585	0.578	0.570	0.563	0.556	0.549	0.542

Table 13 — Rakes

Tillage Tool/Soil Type	Draft, kN/m	Typical Speed, km/h	Drawbar Power, kW/m
Coulter — Chisel Plow			
Fine	8.39	8.1	18.8
Medium	7.30	8.9	17.9
Coarse	5.84	9.7	15.7
Moldboard Plow			
Fine	17.51	7.3	35.2
Medium	13.43	8.1	30.0
Coarse	8.76	8.1	19.6
Field Cultivator			
Fine	5.69	8.1	12.7
Medium	4.96	8.9	12.2
Coarse	3.94	9.7	10.6
Tandem Disk Harrrow			
Fine	5.84	7.3	11.7
Medium	4.96	8.1	11.1
Coarse	4.38	8.9	10.8
Offset or Heavy Disk			
Fine	7.66	7.3	15.4
Medium	6.71	8.1	15.0
Coarse	6.28	8.1	14.0
One-Way Disk			
Fine	5.84	7.3	11.7
Medium	4.38	8.1	9.8
Coarse	2.92	8.1	6.5
V-Ripper			
Fine*	35.03	6.5	62.6
Medium*	27.73	6.9	52.7
Coarse*	17.51	7.3	35.2
Draft, lb/shank; Drawbar power, hp/shank			

Table 14 — Soil Resistance

Compare:	System 1	System 2
	Conventional Tillage	Reduced Tillage
Operation	Liters per Hectare	
Disk Stalks	4.21	4.21
Plow	15.71 (Chisel)	10.29
Disk	6.08	—
Pre-Emergence Spray	0.94	0.94
Field Cultivation	5.61	5.61
Plant	4.68	3.74
Cultivate	4.21	4.21
Combine	14.96	14.96
Total	62.46	43.95

Table 15 — Comparing Tillage Methods

Engine (Fuel Type)	Average Fuel Consumption (Liters per Hour per Rated PTO kW)	Typical Kilograms per Liter
Gasoline	0.345	0.73
Diesel	0.223	0.82
LP-Gas	0.406	0.50–0.53

Table 16 — Average Fuel Consumption and Fuel Weight for Different Types of Tractor Engines

Operation	Energy Required, PTO kW-Hr per Hectare	Liters per Hectare			Operation	Energy Required, PTO kW-Hs per Hectare	Liters per Hectare		
		Gasoline	Diesel	LP-Gas			Gasoline	Diesel	LP-Gas
Shred Stalks	19.3	9.3	6.7	11.2	Combine, Corn and Grain Sorghum	32.4	20.9	15	25.1
Plow 8 Inches Deep	44.9	22	15.7	26.4	Corn Picker	23.2	15	10.7	18
Heavy Offset Disk	25.4	12.4	8.9	15	Mower (Cutterbar)	6.4	4.6	3.3	5.5
Chisel Plow	29.5	14.4	10.3	17.3	Mower Conditioner	13.3	7.9	5.6	9.4
Tandem Disk, Stalks	11.0	5.9	4.2	7.1	Swather	12.2	7.2	5.1	8.6
Tandem Disk, Chiseled	12.7	7.2	5.1	8.6	Rake, Single	4.6	3.3	2.3	3.9
Tandem Disk, Plowed	17.3	8.5	6.1	10.2	Rake, Tandem	2.8	2	1.4	2.3
Field Cultivate	14.7	7.9	5.6	9.4	Baler	9.2	5.9	4.2	7.1
Spring-Tooth Harrow	9.6	5.2	3.7	6.3	Stack Wagon	11.0	6.5	4.7	7.9
Spike-Tooth Harrow	6.3	3.9	2.8	4.7	Sprayer	1.8	1.3	0.9	1.6
Rod Weeder	7.4	3.9	2.8	4.7	Rotary Mower	17.6	10.5	7.5	12.5
Sweep Plow	16.0	7.9	5.6	9.4	Haul Small Grains	11.0	7.9	5.6	9.4
Cultivate Row Crops	11.0	5.9	4.2	7.1	Grain Drying	154.6	78.5	56	94.2
Rolling Cultivator	7.2	4.6	3.3	5.5	Forage Harvester, Green Forage	22.8	12.4	8.9	15
Rotary Hoe	5.2	3.3	2.3	3.9	Forage Harvester, Haylage	30.0	16.4	11.7	19.6
Anhydrous Applicator	17.3	8.5	6.1	10.2	Forage Harvester, Corn Silage	86.0	47.1	33.6	56.5
Planting Row Crops	12.3	6.5	4.7	7.9	Forage Blower, Green Forage	8.5	4.6	3.3	5.5
No-Till Planter	7.2	4.6	3.3	5.5	Forage Blower, Haylage	6.1	3.3	2.3	3.9
Till Plant (With Sweep)	8.3	5.2	3.7	6.3	Forage Blower, Corn Silage	33.5	18.3	13.1	22
Grain Drill	8.7	4.6	3.3	5.5	Forage Blower, High-Moisture Ear Corn	10.9	5.9	4.2	7.1
Combine (Small Grains)	20.3	13.1	9.3	15.7	Haul Forage, Corn Silage	7.4	3.9	2.8	4.7
Combine, Beans	22.1	14.4	10.3	17.3					

Table 17 — Average Energy and Fuel Requirements

	Worksheet For Comparing Alternatives			
Options	1	2	3	4
Machine:	Repair and Keep Using			
a. Age at Purchase, Years				
b. Current Age, Years				
c. Ending Age, Years				
d. Use, Acres/Year				
e. Use, Hectares/Year				
f. Original List Price				
g. "As-Is" Value (Chapter 6, Table 1)				
h. "Cash Boot" Difference				
i. Capital Value (Line g of Option 1 + Line h of Each Option)				
j. Ending "As-Is" Value				
k. Loss in Capital Value				
l. Immediate Repairs				
m. Future Repairs				
n. Finance Charges on Investment Capital				
o. Taxes, Shelter, and Insurance Costs				
p. Total (Lines k + l + m + n + o)				
q. Annual Costs, $/Year (Line p/[Line c – Line b])				
r. Adjustments, $/Year				
Labor, $/Year				
Fuel and Lubricants, $/Year				
Age Penalty, $/Year				
Income Tax Credit[a], $/Year				
s. Total Cost, $/Year (Lines q + r)				
t. Total Cost, $/Acre (Line s/d)				

a. For income tax credit, use the annual average loss in capital value for the depreciation of line i, plus lines l, m, n, and r (labor, fuel, and lubrication costs). Depreciable portion of line i is normally the "cash boot" difference.

Fig. 1 — Worksheet for Comparing Alternatives

Comparison of Four Alternatives

Options Machine:	1 Repair and Keep Using	2 Trade for New	3 Trade for Used	4 Rent for $140.00/Hour
a. Age at Purchase, Years	0	0	4	----
b. Current Age, Years	8	0	4	----
c. Ending Age, Years	18	10	14	----
d. Use, Acres/Year	800	800	800	800
e. Use, Hectares/Year	110	110	110	110
f. Original List Price	$100,000	$240,000	$200,000	$240,000
g. "As-Is" Value (Chapter 6, Table 1)	$40,841	$204,000	$124,327	----
h. "Cash Boot" Difference	----	$163,159	$83,159	----
i. Capital Value (Line g of Option 1 + Line h of Each Option)	$40,841	$204,000	$124,000	
j. Ending "As-Is" Value	$21,998	$85,560	$56,630	----
k. Loss in Capital Value	$18,843	$118,440	$67,640	----
l. Immediate Repairs	$7,000	$0	$0	----
m. Future Repairs	$13,670	$6,180	$10,980	----
n. Finance Charges on Investment Capital	$42,261	$159,258	$99,198	----
o. Taxes, Shelter, and Insurance Costs	$12,568	$57,910	$36,070	
p. Total (Lines k + l + m + n + o)	$94,343	$342,418	$213,888	----
q. Annual Costs, $/Year (Line p/[Line c − Line b])	$9,434	$34,242	$21,388	$15,400
r. Adjustments, $/Year				
Labor, $/Year	$1,100	$1,100	$1,100	$1,100
Fuel and Lubricants, $/Year	$660	$660	$660	$660
Age Penalty, $/Year	$1,500	$0	$750	$0
Income Tax Credit[a], $/Year	($2,607)	($6,161)	($4,255)	($4,189)
s. Total Cost, $/Year (Lines q + r)	$10,087	$29,841	$19,643	$12,971
t. Total Cost, $/Acre (Line s/d)	$12.61	$37.30	$24.55	$16.21

a. For income tax credit, use the annual average loss in capital value for the depreciation of line i, plus lines l, m, n, and r (labor, fuel, and lubrication costs). Depreciable portion of line i is normally the "cash boot" difference.

Fig. 2 — Comparison of Four Alternatives

Explanation of Worksheet

The spreadsheet on page 11-20 has been set up to compare four options. In this explanation actual numbers are used to illustrate the procedure and formulas used.

Lines a, b, c, d, e, and f are estimated and manually entered.

Line g. The "as-is" value can be the actual known value, estimated from Table 1 in Chapter 6 or calculated from formulas given in Chapter 6. "As-is" value for Option 1, the old combine, is estimated to be $40,841. The "as-is" value for Option 2, the new combine, is estimated to be 85% of its list price, or $119,000 (0.85 x $140,000). The "as-is" value for Option 3 is estimated using the formula:

$$RV = LP \times RV1 \times RV2^Y \times 1.2$$

Where: RV = List price, RV1 and RV2 = constants from Chapter 6, and 1.2 provides for 20% reconditioning charge.

For Option 3, the "as-is" value is:

$$RV = \$120{,}000 \times 0.67 \times 0.94^4 \times 1.2 = \$75{,}327$$

Line h. "Cash boot" difference is the amount of capital required to trade for a new combine. For Option 2, the "cash boot" difference is $78,159 ($119,000 – $40,841), and for Option 3, it is $34,486 ($75,327 – $40,841).

Line i. Capital value is the total amount of capital invested. In most cases the capital value is equal to the purchase price of the machine or "as-is" value.

Line j. Ending "as-is" value is estimated using line f and a value from Table 1, Chapter 6, or the remaining value formula in Chapter 6 as used in line g above. For Option 1, the ending "as-is" value is calculated for an ending age of 18 years as follows:

$$RV = \$100{,}000 \times 0.67 \times 0.94^{18} = \$21{,}998$$

The same procedure is used for Option 2 with years equal to 10, and for Option 3 with years equal to 14.

Line k. Loss in capital value is the depreciation over the ownership period, line i – line j. In Option 2, depreciation is equal to $68,478 ($119,000 – $50,522).

Line l. Immediate repair cost is the best estimate of needed repair costs. For Option 1, the combine will be repaired and owned for 10 more years, so it is estimated that $7,000 is needed for immediate repairs.

Line m. Future repair costs are estimated using the repair cost formula from Chapter 8. For combines, the total accumulated repair (TAR) cost formula is:

$$TAR = \text{List Price} \times 0.03982 \times (\text{hours}/1000)^{2.1}$$

For Option 1, future repair cost is equal to TAR for 1,980 hours (18 years x 110 hours per year) minus TAR for 880 hours (8 years x 110 hours per year).

For Option 2, future repair cost is equal to TAR calculated for 1,100 hours (10 years x 110 hours per year).

For Option 3, future repair cost is equal to TAR for 1,540 hours (14 years x 110 hours per year) minus TAR for 440 hours (4 years x 110 hours per year).

Line n. Finance charges apply to the capital value of the combine plus immediate repair cost. Using a short cut method, the finance charge on capital value is equal to the average value times the interest rate times the number of years. An interest rate of 11% is used in this example.

For Option 1, finance charge would be:

$$\left(\frac{\$40{,}841 + \$21{,}998}{2}\right) + \$7{,}000 \times 0.11 \times 10 = \$42{,}261$$

For Option 2, finance charge would be:

$$\left(\frac{\$119{,}000 + \$50{,}522}{2}\right) \times 0.11 \times 10 = \$93{,}237$$

For Option 3, the same procedure is used.

Line o. Taxes, shelter, and insurance cost (TSI) is equal to 4% of the average value times the number of years:

$$TSI = 0.04 \times \left(\frac{\text{Line i} + \text{Line j}}{2}\right) \times (\text{Line c} - \text{Line b})$$

For Option 1,

$$TSI = 0.04 \times \left(\frac{\$40{,}841 + \$21{,}998}{2}\right) \times (18 - 8)$$

Line p. Total cost equals the sum of lines k, l, m, n, and o.

Line q. Average annual cost in dollars per year is equal to:

$$\frac{\text{Line p}}{\text{Line c} - \text{Line b}}$$

Line r. Adjustments to annual cost for labor and for fuel and lubricants are optional. However, if these costs are included in one option, they should be included in all options.

An age penalty is any loss in crop quality or quantity expected because of the age of a machine. In Option 1, an age penalty of $1,500 per year is estimated for the old combine, due to expected high harvest losses.

Income tax credits are applied to lines k, l, m, n, and o, and to costs for fuel, lubricants, and labor in line r. A marginal income tax rate of 28% is assumed. For Option 1, it is assumed depreciation has already been used for income tax purposes and, therefore, the deduction for Option 1 is:

$$0.28 \times \left(\frac{\$7{,}000 + \$13{,}670 + \$42{,}261 + \$12{,}568}{10} \right) + (\$1{,}100 + \$660)$$

= $2,607/yr

Lines s and t. Total cost is $10,087 per year or $12.61 per acre for Option 1, repair and keep using the old combine. For the new combine, the cost is $15,842 per year or $19.80 per acre. The savings Repairing and continuing to use the old combine is $5,755 per year or $7.19 per acre.

This comparison illustrates, as is often the case, that it is less expensive to repair and keep using an old machine. Of course, there are limits to how long a machine should be used. Reliability may become a major factor in using the old machine for a total of 20 years, making it more economical to trade for a new machine. Reliability is not accounted for in the comparison.

Nebraska OECD Tractor Test 1775 — Summary 310

John Deere 8310 Diesel, 16 Speed

Location of Test: Nebraska Tractor Test Laboratory, University of Nebraska, Lincoln, Nebraska 68583-0832

Dates of Test: April 13–May 12, 2000

Manufacturer: John Deere Waterloo Works, P.O. Box 270, Waterloo, Iowa, USA 50704

FUEL, OIL and TIME: Fuel No. 2 Diesel, **Specific gravity converted to 60°/60°F (15°/15°C)** 0.8487 **Fuel weight** 7.067 lb/gal (0.847 kg/L) **Oil SAE** 15W-40 **API service classification** CF-4 **Transmission and hydraulic lubricant** John Deere Hy-Gard fluid **Front axle lubricant** SAE 85W-140 API GL-5 **Total time engine was operated:** 29.0 hours

ENGINE: Make John Deere Diesel **Type** six cylinder vertical with turbocharger and air-to-air aftercooler **Serial No.** *RG6081H098910* **Crankshaft** lengthwise **Rated engine speed** 2200 **Bore and stroke** 4.56" x 5.06" (115.8 mm x 128.5 mm) **Compression ratio** 16.5 to 1 **Displacement** 496 cu in. (8134 mL) **Starting system** 12 volt **Lubrication** pressure **Air cleaner** two paper elements and aspirator **Oil filter** one full flow cartridge **Oil cooler** engine coolant heat exchanger for crankcase oil, radiator for hydraulic and transmission oil **Fuel filter** one paper element and prestrainer **Fuel cooler** radiator for pump return fuel **Muffler** vertical **Cooling medium temperature control** 2 thermostats and variable speed fan

ENGINE OPERATING PRAMAETERS: Fuel rate: 79.9–87.4 lb/h (36.3–39.7 kg/h) **High idle:** 2275–2325 rpm **Turbo boost:** nominal 18.7–23.1 psi (129–159 kPa) as measured 21.3 psi (147 kPa)

CHASSIS: Type front wheel assist **Serial No.** *RW8310P001626* **Tread width** rear 65.3" (1659 mm) to 141.1" (3585 mm) front 60.0" (1524 mm) to 88.0" (2235 mm) **Wheelbase** 116.1" (2950 mm) **Hydraulic control system** direct engine drive **Transmission** selective gear fixed ratio with full range operator controlled power shift **Nominal travel speeds mph (km/h)** first 1.35 (2.18) second 1.73 (2.79) third 2.21 (3.55) fourth 2.81 (4.53) fifth 3.41 (5.49) sixth 3.85 (6.19) seventh 4.36 (7.01) eighth 4.92 (7.91) ninth 5.54 (8.92) tenth 6.62 (10.07) eleventh 7.08 (11.40) twelfth 7.99 (12.86) thirteenth 10.17 (16.36) fourteenth 12.99 (20.91) fifteenth 16.53 (26.61) sixteenth 23.04 (37.08) @ 2400 engine rpm reverse 1.18 (1.90), 2.98 (4.79). 3.66 (5.89), 6.45 (10.38) @ 1600 engine rpm **Clutch** wet multiple disc hydraulically activated by foot pedal **Brakes** wet multiple disk hydraulically operated by two foot pedals that can be locked together **Steering** hydrostatic **Power take-off** 1-3/4" shaft — 1000 rpm at 2179 engine rpm (optional — 1-3/8" shaft, 540 rpm at 1978 engine rpm or 1000 rpm at 2179 engine rpm) **Unladen tractor mass** 19860 lb (9008 kg)

REPAIRS AND ADJUSTMENTS: No repairs or adjustments.

REMARKS: All test results were determined from observed data obtained in accordance with official OECD, SAE and Nebraska test procedures. For the maximum power tests the fuel temperature at the injection pump inlet was maintained at 115°F (46°C). The performance figures on this summary were taken from a test conducted under the OECD Code II test code procedure.

We, the undersigned, certify that this is a true and correct report of the official Tractor Test No. 1775, Nebraska Summary 310, July 7, 2000.

Brent T. Sampson
Test Engineer

L. L. Bashford
M. F. Kocher
R. D. Grisso, Jr.
Board of Tractor Test Engineers

POWER TAKE-OFF PERFORMANCE

Power HP (kW)	Crank shaft speed rpm	Gal/hr (l/h)	lb/hp.hr (kg/kW.h)	Hp.hr/gal (kW.h/l)	Mean Atmospheric Conditions
MAXIMUM POWER AND FUEL CONSUMPTION					
Rated Engine Speed—(PTO speed—1009 rpm)					
207.25 (154.55)	2200	11.63 (44.01)	0.396 (0.241)	17.82 (3.51)	
Maximum Power (2 hours)					
236.74 (176.54)	2000	12.43 (47.04)	0.371 (0.226)	19.05 (3.75)	
VARYING POWER AND FUEL CONSUMPTION					
207.25 (154.55)	2200	11.63 (44.01)	0.396 (0.241)	17.82 (3.51)	Air temperature
180.63 (134.70)	2254	10.57 (40.01)	0.414 (0.252)	17.09 (3.37)	76°F (24°C)
136.00 (101.42)	2265	8.58 (32.46)	0.446 (0.271)	15.86 (3.12)	Relative humidity
90.77 (67.68)	2273	6.62 (25.07)	0.516 (0.314)	13.71 (2.70)	48%
45.89 (34.22)	2286	4.58 (17.35)	0.706 (0.429)	10.01 (1.97)	Barometer
1.00 (0.75)	2293	2.72 (10.28)	19.171 (11.661)	0.37 (0.07)	28.91" Hg (97.90 kPa)

Maximum Torque - 741 lb.-ft. *(1005 Nm)* at 1100 rpm
Maximum Torque Rise - 49.8%
Torque rise at 1800 engine rpm - 36%

DRAWBAR PERFORMANCE
UNBALLASTED - FRONT DRIVE ENGAGED
FUEL CONSUMPTION CHARACTERISTICS

Power Hp (kW)	Drawbar pull lbs (kN)	Speed mph (km/h)	Crank-shaft speed rpm	Slip %	Fuel Consumption lb/hp.hr (kg/kW.h)	Fuel Consumption Hp.hr/gal (kW.h/l)	Temp.°F (°C) cooling med	Air dry bulb	Barom. inch Hg (kPa)
Maximum Power—8th Gear									
176.22 (131.41)	14149 (62.94)	4.67 (7.52)	2200	6.26	0.467 (0.284)	15.12 (2.98)	195 (90)	64 (18)	28.96 (98.07)
75% of Pull at Maximum Power—8th Gear									
139.21 (103.81)	10659 (47.41)	4.90 (7.88)	2259	4.31	0.490 (0.298)	14.41 (2.84)	191 (88)	64 (18)	28.98 (98.14)
50% of Pull at Maximum Power—8th Gear									
94.47 (70.45)	7096 (31.56)	4.99 (8.04)	2271	2.91	0.561 (0.341)	12.60 (2.48)	184 (84)	65 (18)	28.99 (98.17)
75% of Pull at Reduced Engine Speed—10th Gear									
139.16 (103.77)	10672 (47.47)	4.89 (7.87)	1772	4.31	0.425 (0.258)	16.63 (3.28)	194 (90)	64 (18)	28.98 (98.14)
50% of Pull at Reduced Engine Speed—10th Gear									
94.36 (70.36)	7106 (31.61)	4.98 (8.01)	1780	2.91	0.468 (0.285)	15.10 (2.97)	188 (86)	65 (18)	28.99 (98.17)

DRAWBAR PERFORMANCE
UNBALLASTED - FRONT DRIVE ENGAGED
MAXIMUM POWER IN SELECTED GEARS

Power Hp (kW)	Drawbar pull lbs (kN)	Speed mph (km/h)	Crank-shaft speed rpm	Slip %	Fuel Consumption lb/hp.hr (kg/kW.h)	Hp.hr/gal (kW.h/l)	Temp.°F (°C) cooling med	Air dry bulb	Barom. inch Hg (kPa)
				5th Gear					
149.29 (111.33)	18508 (82.33)	3.03 (4.87)	2253	14.50	0.522 (0.318)	13.54 (2.67)	193 (89)	59 (15)	28.96 (98.07)
				6th Gear					
163.43 (121.87)	17622 (78.38)	3.48 (5.60)	2244	12.60	0.506 (0.308)	13.98 (2.75)	192 (89)	61 (16)	28.96 (98.07)
				7th Gear					
176.48 (131.60)	17105 (76.09)	3.87 (6.23)	2159	10.84	0.478 (0.291)	14.78 (2.91)	196 (91)	64 (18)	28.96 (98.07)
				8th Gear					
188.92 (140.88)	16946 (75.38)	4.18 (6.73)	2066	10.61	0.462 (0.281)	15.31 (3.02)	196 (91)	64 (18)	28.96 (98.07)
				9th Gear					
198.91 (148.33)	15960 (70.99)	4.67 (7.52)	1997	8.61	0.443 (0.269)	15.96 (3.14)	198 (92)	64 (18)	28.97 (98.10)
				10th Gear					
202.42 (150.94)	14083 (62.64)	5.39 (8.67)	1999	6.43	0.439 (0.267)	16.11 (3.17)	198 (92)	64 (18)	28.97 (98.10)
				11th Gear					
203.56 (151.80)	12339 (54.88)	6.19 (9.96)	1999	5.08	0.428 (0.261)	16.49 (3.25)	199 (93)	66 (19)	28.97 (98.10)
				12th Gear					
203.01 (151.38)	10824 (48.15)	7.03 (11.32)	1998	4.40	0.432 (0.263)	16.37 (3.22)	198 (92)	64 (18)	28.98 (98.13)
				13th Gear					
200.95 (149.85)	8314 (36.98)	9.06 (14.59)	2000	3.17	0.436 (0.265)	16.21 (3.19)	201 (94)	64 (18)	28.98 (98.13)

TRACTOR SOUND LEVEL WITH CAB	Front Wheel Drive Engaged dB(A)	Disengaged dB(A)
At no load in 8th gear	73.5	73.5
Transport speed - no load - 16th gear		75.8
Bystander in 16th Gear		88.6

TIRES, BALLAST AND WEIGHT	With Ballast	Without Ballast
Rear Tires - No., size, ply & psi(kPa)	Four 620/70R42;**;9(60)	Two 620/70R42;**;13(90)
Ballast - Duals (total)	2310 lb (1048 kg)	None
- Cast Iron (total)	3075 lb (1395 kg)	None
Front Tires - No., size, ply & psi(kPa)	Two 480/70R30;***;25(170)	Two 480/70R30;***;16(110)
Ballast - Liquid (total)	None	None
- Cast Iron (total)	1750 lb (794 kg)	None
Height of Drawbar	20.5 in (520 mm)	21.0 in (535 mm)
Static Weight with operator - Rear	16715 lb (7582 kg)	11905 lb (5400 kg)
- Front	10450 lb (4740 kg)	8125 lb (3685 kg)
- Total	27165 lb (12322 kg)	20030 lb (9085 kg)

DRAWBAR PERFORMANCE
BALLASTED- FRONT DRIVE ENGAGED (2000 RPM)
MAXIMUM POWER IN SELECTED GEARS

Power Hp (kW)	Drawbar pull lbs (kN)	Speed mph (km/h)	Crank-shaft speed rpm	Slip %	Fuel Consumption lb/hp.hr (kg/kW.h)	Hp.hr/gal (kW.h/l)	Temp.°F(°C) cooling med	Air dry bulb	Barom. inch Hg (kPa)
					3rd Gear				
136.87 (102.06)	26601 (118.32)	1.93 (3.11)	2255	14.94	0.537 (0.327)	13.15 (2.59)	187 (86)	57 (14)	29.01 (98.24)
					4th Gear				
165.87 (123.69)	24858 (110.57)	2.50 (4.03)	2188	11.21	0.496 (0.302)	14.24 (2.81)	191 (88)	58 (14)	29.02 (98.27)
					5th Gear				
188.10 (140.26)	24417 (108.61)	2.89 (4.65)	2068	10.32	0.459 (0.279)	15.40 (3.03)	193 (89)	60 (16)	29.02 (98.27)
					6th Gear				
197.30 (147.13)	22942 (102.05)	3.23 (5.19)	1998	8.32	0.444 (0.270)	15.93 (3.14)	197 (92)	64 (18)	29.01 (98.24)
					7th Gear				
204.17 (152.25)	20419 (90.83)	3.75 (6.03)	2001	5.90	0.430 (0.261)	16.45 (3.24)	198 (92)	65 (18)	29.02 (98.27)
					8th Gear				
205.52 (153.26)	18009 (80.11)	4.28 (6.89)	2000	4.55	0.427 (0.259)	16.57 (3.26)	199 (93)	66 (19)	29.02 (98.27)
					9th Gear				
206.74 (154.17)	15933 (70.87)	4.87 (7.83)	2000	3.95	0.425 (0.259)	16.63 (3.28)	199 (93)	68 (20)	29.01 (98.24)
					10th Gear				
206.62 (154.08)	14030 (62.41)	5.52 (8.89)	2000	3.43	0.423 (0.258)	16.69 (3.29)	200 (93)	71 (22)	28.99 (98.17)
					11th Gear				
205.21 (153.03)	12260 (54.54)	6.28 (10.10)	1999	2.91	0.428 (0.260)	16.51 (3.25)	195 (91)	73 (23)	28.97 (98.10)
					12th Gear				
203.44 (151.71)	10730 (47.73)	7.11 (11.44)	1998	2.55	0.432 (0.263)	16.37 (3.22)	198 (92)	73 (23)	28.97 (98.10)
					13th Gear				
198.58 (148.08)	8162 (36.31)	9.12 (14.68)	2002	2.11	0.442 (0.269)	16.01 (3.15)	198 (92)	73 (18)	28.96 (98.07)

DRAWBAR PERFORMANCE
BALLASTED-FRONT DRIVE ENGAGED (2200 RPM)
MAXIMUM POWER IN SELECTED GEARS

Power Hp (kW)	Drawbar pull lbs (kN)	Speed mph (km/h)	Crank-shaft speed rpm	Slip %	Fuel Consumption lb/hp.hr (kg/kW.h)	Hp.hr/gal (kW.h/l)	Temp.°F(°C) cooling med	Air dry bulb	Barom. inch Hg (kPa)
					3rd Gear				
135.36 (100.94)	26123 (116.20)	1.94 (3.13)	2256	14.34	0.537 (0.327)	13.15 (2.59)	186 (86)	57 (14)	29.01 (98.24)
					4th Gear				
167.85 (125.16)	24475 (108.87)	2.57 (4.14)	2201	9.10	0.487 (0.296)	14.51 (2.86)	192 (89)	57 (14)	29.01 (98.24)
					5th Gear				
177.90 (132.66)	20655 (91.88)	3.23 (5.20)	2201	5.90	0.461 (0.280)	15.33 (3.02)	194 (90)	62 (17)	29.01 (98.24)
					6th Gear				
177.74 (132.54)	18114 (80.58)	3.68 (5.92)	2198	4.72	0.462 (0.281)	15.28 (3.01)	194 (90)	64 (18)	29.01 (98.24)
					7th Gear				
181.38 (135.25)	16213 (72.12)	4.20 (6.75)	2197	4.13	0.453 (0.276)	15.60 (3.07)	195 (90)	64 (18)	29.02 (98.27)
					8th Gear				
180.46 (134.57)	14183 (63.09)	4.77 (7.68)	2201	3.35	0.455 (0.277)	15.53 (3.06)	197 (91)	67 (19)	29.02 (98.27)
					9th Gear				
179.88 (134.14)	12479 (55.51)	5.41 (8.70)	2200	2.99	0.459 (0.279)	15.41 (3.04)	197 (92)	69 (21)	29.00 (98.21)
					10th Gear				
179.10 (133.55)	10969 (48.79)	6.12 (9.85)	2200	2.46	0.460 (0.280)	15.35 (3.02)	195 (91)	72 (22)	28.98 (99.13)
					11th Gear				
175.26 (130.70)	9444 (42.01)	6.96 (11.20)	2200	2.20	0.468 (0.284)	15.11 (2.98)	196 (91)	73 (23)	28.97 (98.10)
					12th Gear				
173.90 (129.68)	8296 (36.90)	7.86 (12.65)	2196	1.93	0.475 (0.289)	14.88 (2.93)	197 (91)	73 (23)	28.96 (98.07)

THREE POINT HITCH PERFORMANCE (OECD Static Test)

CATEGORY: III
Quick Attach: Yes

	lift cylinders 2x90 mm	lift cylinders 2x100 mm
Maximum Force Exerted Through Whole Range:	10827 lbs (48.2 kN)	15147 lbs (67.4 kN)
i) Opening pressure of relief valve:	NA	NA
		High flow option
Sustained pressure at compensator cutoff:	2920 psi (201 bar)	2910 psi (201 bar)
	two outlet sets combined	
ii) Pump delivery rate at minimum pressure and rated engine speed:	35.1 GPM (132.9 l/min)	44.3 GPM (167.7 l/min)
iii) Pump delivery rate at maximum hydraulic power:	33.2 GPM (125.7 l/min)	41.6 GPM (157.5 l/min)
Delivery pressure:	2550 psi (176 bar)	2460 psi (170 bar)
Power:	49.4 HP (36.8 kW)	59.7 HP (44.5 kW)
	single outlet set	
ii) Pump delivery rate at minimum pressure and rated engine speed:	31.8 GPM (120.4 l/min)	32.9 GPM (124.5 l/min)
iii) Pump delivery rate at maximum hydraulic power:	30.4 GPM (115.1 l/min)	29.8 GPM (112.8 l/min)
Delivery pressure:	2250 psi (155 bar)	2230 psi (154 bar)
Power:	39.9 HP (29.8 kW)	38.8 HP (28.9 kW)

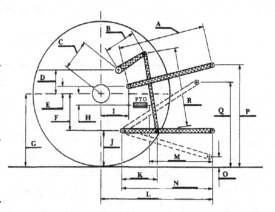

THREE POINT HITCH PERFORMANCE

Observed Maximum Pressure psi. (bar)	2920 (201)
Location:	remote outlet
Hydraulic oil temperature: °F (°C)	150 (65)
Location:	pump inlet
Category:	III
Quick attach:	yes

SAE Static Test—System pressure 2640 psi (182 Bar)
with lift cylinders (2) 90 mm

Hitch point distance to ground level in. (mm)	8.0 (203)	16.0 (406)	24.0 (610)	32.0 (813)	40.0 (1016)
Lift force on frame lb	11637	12371	11957	11642	10755
" " " " " " (kN)	(51.8)	(55.0)	(53.2)	(51.8)	(47.8)

with lift cylinders (2) 100 mm

Hitch point distance to ground level in. (mm)	8.0 (203)	16.0 (406)	24.0 (610)	32.0 (813)	40.0 (1016)
Lift force on frame lb	16799	17262	16754	16430	15260
" " " " " " (kN)	(74.7)	(76.8)	(74.5)	(73.1)	(67.9)

ASAE Static Test—System pressure 2900 psi (200 Bar)
with lift cylinders (2) 90 mm

Hitch point distance to ground level in. (mm)	8.0 (203)	16.0 (406)	24.0 (610)	32.0 (813)	40.0 (1016)
Lift force on frame lb	12821	13643	13196	12820	11852
" " " " " " (kN)	(57.0)	(60.7)	(58.7)	(57.0)	(52.7)

with lift cylinders (2) 100 mm

Hitch point distance to ground level in. (mm)	8.0 (203)	16.0 (406)	24.0 (610)	32.0 (813)	40.0 (1016)
Lift force on frame lb	18591	19018	18376	18070	16742
" " " " " " (kN)	(82.7)	(84.6)	(81.7)	(80.4)	(74.5)

HITCH DIMENSIONS AS TESTED—NO LOAD

	inch	mm
A	28.3	718
B	19.5	495
C	21.7	550
D	19.5	495
E	4.8	123
F	13.8	350
G	35.6	905
H	7.8	197
I	20.3	515
J	21.8	555
K	28.2	716
L	48.9	1242
*L'	52.4	1331
M	22.0	558
N	38.1	967
O	9.0	229
P	43.8	1114
Q	40.1	1019
R	41.5	1054

*L' to Quick Attach ends

Parameters for Estimating Repair Costs

Proposed values for wear-out life and repair and maintenance cost parameters for agricultural equipment. This information was used as a basis for the repair cost parameters in Chapter 8[1].

Machine	Wear-Out Life, Hours	Total Life, Repair, Costs, %	Repair Cost Parameters RF1	RF2
Tractors				
Two-Wheel Drive	12,000	100	0.06944	2.0
Four-Wheel Drive and Crawler	16,000	80	0.003125	2.0
Tillage and Planting				
Moldboard Plow	2,000	100	0.28730	1.8
Heavy-Duty Disk	2,000	70	0.26524	1.4
Tandem Disk Harrow	2,000	70	0.26524	1.4
Chisel Plow	2,000	70	0.26524	1.4
Field Cultivator	2,000	70	0.26524	1.4
Spring-Tooth Harrow	2,000	70	0.26524	1.4
Roller/Mulcher-Packer	2,000	70	0.26524	1.4
Rotary Hoe	2,000	70	0.26524	1.4
Row Crop Cultivator	2,000	70	0.26524	1.4
Rotary Tiller	1,500	80	0.36000	2.0
Row Crop Planter	1,500	75	0.32000	2.1
Grain Drill	1,500	75	0.32000	2.1
Harvesting				
Corn Picker Sheller	2,000	40	0.03920	2.1
Combine	3,000	40	0.03920	2.1
Combine, Self-Propelled	3,000	40	0.03920	2.1
Mower	2,000	150	0.46170	1.7
Mower, Rotary	2,000	150	0.46170	1.7
Mower-Conditioner	2,000	150	0.46170	1.7
Mower-Conditioner, Rotary	2,000	150	0.46170	1.7
Windrower, Self-Propelled	3,000	55	0.16110	2.0
Side Delivery Rake	2,500	60	0.16630	1.4
Rectangular Baler	2,500	75	0.014410	1.8
Large Rectangular Baler	2,500	75	0.014410	1.8
Large Round Baler	1,500	90	0.434000	1.8
Forage Harvester	2,500	65	0.15000	1.6
Forage Harvester, Self-Propelled	4,000	50	0.031250	2.0
Sugar Beet Harvester	1,500	100	0.59000	1.3
Potato Harvester	2,500	70	0.19000	1.4
Cotton Picker or Stripper	5,000	60	0.074044	1.3

Total accumulated repair costs = List price x RF1 x (total hrs./100)^{RF2}
The effect of inflation is not included in these estimates.

Table 18 — Parameters for Estimating Repair Costs

1. Source: American Society of Agricultural Engineers

Weights and Measures

Measurement Conversion Chart

Metric to English

LENGTH
1 millimeter = 0.03937 inch .... in
1 meter = 3.281 feet .... ft
1 kilometer = 0.621 mile .... mi

AREA
1 meter^2 = 10.76 feet^2 .... ft^2
1 hectare = 2.471 acres .... acre
 (hectare = 10,000 m^2)

MASS (WEIGHT)
1 kilogram = 2.205 pounds .... lb
1 tonne (1000 kg) = 1.102 short tons .... sh tn

VOLUME
1 meter^3 = 35.31 feet^3 .... ft^3
1 meter^3 = 1.308 yards^3 .... yd^3
1 meter^3 = 28.38 bushel .... bu
1 liter = 0.02838 bushel .... bu
1 liter = 1.057 quarts .... qt

PRESSURE
1 kilopascal = 0.145 pound/in^2 .... psi

STRESS
1 megapascal or
1 newton/millimeter^2 = 145 pound/in^2 .... psi
 (1N/mm^2 = 1MPa)

POWER
1 kilowatt = 1.341 horsepower (550 lb-ft/s) .... hp
 (1 watt = 1 N•m/sec)

ENERGY (WORK)
1 joule = 0.0009478 British Thermal Unit .... Btu
 (1 J = 1 W*s)

FORCE
1 newton = 0.2248 pound force .... lb force

TORQUE OR BENDING MOMENT
1 newton meter = 0.7376 foot-pound .... lb-ft

TEMPERATURE
$t_C = (t_F - 32)/1.8$

English to Metric

LENGTH
1 inch = 25.4 millimeters .... mm
1 foot = 0.3048 meter .... m
1 yard = 0.9144 meter .... m
1 mile = 1.608 kilometers .... km

AREA
1 foot^2 = 0.0929 meter^2 .... m^2
1 acre = 0.4047 hectare .... ha
 (hectare = 10,000 m^2)

MASS (WEIGHT)
1 pound = 0.4535 kilogram .... kg
1 ton (2000 lb) = 0.9071 tonne .... t

VOLUME
1 foot^3 = 0.02832 meter^3 .... m^3
1 yard^3 = 0.7646 meter^3 .... m^3
1 bushel = 0.03524 meter^3 .... m^3
1 bushel = 35.24 liters L
1 quart = 0.9464 liter L
1 gallon = 3.785 liters L

PRESSURE
1 pound/inch^2 = 6.895 kilopascals .... kPa
1 pound/inch^2 = 0.06895 bar .... bar

STRESS
1 pound/in^2 (psi) = ... 0.006895 megapascal (MPa)
 .... or newton/mm^2 (N/mm^2) (1 N/mm^2 = 1 MPa)

POWER
1 horsepower (550 lb-ft/s) = 0.7457 kilowatt ... kW
 (1 watt = 1 N•m/s)

ENERGY (WORK)
1 British Thermal Unit = 1055 joules .... J
 (1 J = 1 W*s)

FORCE
1 pound = 4.448 newtons .... N

Torque or Bending Moment
1 pound-foot = 1.356 newton-meters .... N•m

TEMPERATURE
$t_F = 1.8 \times t_C + 32$

ACREAGE CHART FOR VARIOUS ROW LENGTHS AND IMPLEMENT WIDTHS

Row Length (Feet)	Width to Equal One Acre (Feet)	Number of Rows to Equal One Acre Row Width, Inches					
		20	24	28	36	38	40
300	145.2	87.0	72.6	62.2	48.5	45.9	43.5
400	108.9	65.2	55.0	46.7	36.3	34.4	32.6
500	87.1	52.2	43.5	37.4	29.0	27.5	26.1
600	72.6	43.6	36.3	31.1	24.2	23.0	21.8
700	62.2	37.4	31.0	26.6	20.7	19.6	18.7
800	54.5	32.6	27.2	23.4	18.2	17.2	16.3
900	48.4	29.0	24.2	20.7	16.1	15.3	14.5
1,000	43.6	26.2	21.8	18.7	14.5	13.8	13.1
1,100	39.6	23.8	19.8	17.0	13.2	12.5	11.9
1,200	36.3	21.8	18.1	15.5	12.1	11.5	10.9
1,300	33.5	20.0	16.7	14.3	11.2	10.6	10.0
1,400	31.1	18.6	15.5	13.3	10.4	9.8	9.3
1,500	29.0	17.4	14.5	12.4	9.7	9.2	8.7
1,600	27.2	16.4	13.6	11.6	9.1	8.6	8.2

Row Length (Feet)	Number of Trips to Equal One Acre Implement Width, Feet							
	4	5	6	8	10	12	14	16
300	36.3	29.0	24.2	18.2	14.5	12.1	10.4	9.1
400	27.2	21.8	18.2	13.6	10.9	9.1	7.8	6.8
500	21.8	17.4	14.5	10.9	8.7	7.3	6.2	5.4
600	18.2	14.5	12.1	9.1	7.3	6.0	5.2	4.5
700	15.6	12.4	10.4	7.8	6.2	5.2	4.4	3.9
800	13.6	10.9	9.1	6.8	5.5	4.5	3.9	3.4
900	12.1	9.7	8.1	6.1	4.8	4.0	3.5	3.0
1,000	10.9	8.7	7.3	5.4	4.4	3.6	3.1	2.7
1,100	9.9	7.9	6.6	4.9	4.0	3.3	2.8	2.5
1,200	9.1	7.3	6.1	4.5	3.6	3.0	2.6	2.3
1,300	8.4	6.7	5.6	4.2	3.4	2.8	2.4	2.1
1,400	7.8	6.2	5.2	3.9	3.1	2.6	2.2	1.9
1,500	7.2	5.8	4.8	3.6	2.9	2.4	2.1	1.8
1,600	6.8	5.4	4.5	3.4	2.7	2.2	1.9	1.7

SEEDS OR PLANTS PER ACRE, THOUSANDS

Row Spacing (Inches)	Seeds or Plant Spacing, Inches										
	1	2	4	6	8	10	12	14	16	18	20
20	313.6	156.8	78.4	52.3	39.2	31.4	26.1	22.4	19.6	17.4	15.7
22	285.1	142.6	71.3	47.5	35.6	28.5	23.8	20.4	17.8	15.8	14.3
24	261.4	130.7	65.3	43.6	32.7	26.1	21.8	18.7	16.3	14.5	13.1
26	241.2	120.6	60.3	40.2	30.2	24.1	20.1	17.2	15.1	13.4	12.1
28	224.0	112.0	56.0	37.3	28.0	22.4	18.7	16.0	14.0	12.4	11.2
30	209.1	104.5	52.3	34.8	26.1	20.9	17.4	14.9	13.1	11.6	10.4
32	196.0	98.0	49.0	32.7	24.5	19.6	16.3	14.0	12.2	10.9	9.8
34	184.5	92.2	46.1	30.7	23.1	18.4	15.4	13.2	11.5	10.2	9.2
36	174.2	87.1	43.6	29.0	21.8	17.4	14.5	12.4	10.9	9.7	8.7
38	165.1	82.5	41.3	27.5	20.6	16.5	13.8	11.8	10.3	9.2	8.2
40	156.8	78.4	39.2	26.1	19.6	15.7	13.1	11.2	9.8	8.7	7.8
42	149.3	74.7	37.3	24.9	18.7	14.9	12.4	10.7	9.3	8.3	7.5

DETERMINING HECTARES FOR VARIOUS ROW LENGTHS AND IMPLEMENT WIDTHS

Row Length (Meters)	Width to Equal One Hectare (Meters)	Number of Rows to Equal One Hectare						
		Row Width, Centimeters						
		51	61	71	76	91	97	102
100	100	196.9	164	140.6	131.2	109.4	103.6	98.4
150	67	131.2	109.4	93.7	87.5	73	69.1	65.6
200	50	98.4	82	70.3	65.6	54.7	51.8	49.2
250	40	78.7	65.5	56.2	52.5	43.7	41.4	39.4
300	33.3	65.6	54.7	46.8	43.7	36.5	34.5	32.8
350	28.6	56.2	46.9	40.2	37.5	31.2	29.6	28.1
400	25	49.2	41	35.2	32.8	27.3	26.0	24.6
450	22.2	43.7	36.5	31.2	29.2	24.3	23.0	21.9
500	20	39.4	32.8	28.1	26.2	21.9	20.7	19.7

Number of Trips to Equal One Hectare

Row Length (Meters)	Implement Width, Meters								
	1	1.5	2	2.5	3	3.5	4	4.5	5
100	100	66.7	50	40	33.3	28.6	25	22.2	20
150	60.7	44.4	33.3	26.7	22.2	19	16.7	14.8	13.3
200	50	33.3	25	20	16.7	14.3	12.5	11.1	10
250	40	26.7	20	16	13.3	11.4	10	8.9	8
300	33.3	22.2	16.7	13.3	11.1	9.5	8.3	7.4	6.7
350	28.6	19.0	14.3	11.4	9.5	8.2	7.1	6.3	5.7
400	25	16.7	12.5	10	8.3	7.1	6.25	5.6	5
450	22.2	14.8	11.1	8.9	7.4	6.3	5.6	4.9	4.4
500	20	13.3	10	8	6.7	5.7	5	4.4	4

SEEDS OR PLANTS PER HECTARE, THOUSANDS

Row Spacing (Centimeters)	Seeds or Plant Spacing, Centimeters									
	2.5	5	10	15	20	25	30	35	40	45	50
51	787.6	393.8	196.9	131.3	98.5	78.8	65.6	56.3	49.2	43.7	39.4
61	656	328	164	109.4	82	65.6	54.6	46.9	41	36.4	32.8
71	562.4	281.2	140.6	93.8	70.3	56.2	46.8	40.2	35.2	31.2	28.1
76	524.8	262.4	131.2	87.5	65.6	52.5	43.7	37.5	32.8	29.1	26.2
91	437.6	218.8	109.4	73	54.7	43.8	36.4	31.3	27.4	24.3	21.9
97	414.4	207.2	103.6	69.1	51.8	41.4	34.5	29.6	25.9	23	20.7
102	393.6	196.8	98.4	65.6	49.2	39.4	32.8	28.1	24.6	21.8	19.7

Probabilities for a Working Day

Region		Central Illinois		State of Iowa		Southeastern Michigan		State of South Carolina		Southern Ontario Canada		Mississippi Delta	
Soil		Prairie soils		State average		Clay loam		Clay loam		Clay loam		Clay	
Notes		18 yrs. data		17 yrs. data		Simulation (tillage only)		Simulation (tillage only)		Simulation (tillage only)		Simulation (tillage only)	
		In early spring and late fall, PWD in Iowa and Illinois may be 0.07 greater in north and west and 0.07 less in south and east						Sandy soils can be worked all months and have higher PWD		Start 7–10 days earlier on sandy soils, 0.15 greater PWD		Non-tillage field work PWD and PWD for sandy soils some greater in winter and early spring	
Average Date	Biweekly Period	Probability level, %											
		50	90	50	90	50	90	50	90	50	90	50	90
Jan. and Feb.	—	0.0	0.0	0.0	0.0	0.0	0.0	0.1	0.0	0.0	0.0	0.07	0.0
Mar. 7	1	0.0	0.0	0.0	0.0	0.0	0.0	—	—	0.0	0.0	—	—
Mar. 21	2	0.29	0.0	0.0	0.0	0.0	0.0	0.03	0.0	0.0	0.0	0.18	0.0
Apr. 4	3	0.42	0.13	0.39	0.16	0.0	0.0	—	—	0.01	0.0	—	—
Apr. 18	4	0.47	0.19	0.57	0.38	0.20	0.0	0.29	0.06	0.07	0.0	0.35	0.08
May 2	5	0.54	0.31	0.66	0.48	—	—	—	—	0.62	0.02	—	—
May 16	6	0.61	0.34	0.68	0.47	0.61	0.32	0.64	0.37	0.60	0.02	0.58	0.28
May 30	7	0.63	0.40	0.66	0.47	—	—	—	—	0.79	0.16	—	—
June 13	8	0.66	0.41	0.69	0.52	0.69	0.42	0.72	0.46	0.77	0.22	0.69	0.39
June 27	9	0.72	0.53	0.74	0.57	—	—	—	—	0.80	0.23	—	—
July 11	10	0.72	0.52	0.77	0.64	0.75	0.52	0.67	0.43	—	—	0.63	0.25
July 25	11	0.72	0.54	0.80	0.67	—	—	—	—	—	—	—	—
Aug. 8	12	0.78	0.64	0.80	0.68	0.74	0.53	0.73	0.51	—	—	0.72	0.45
Aug. 22	13	0.86	0.74	0.86	0.79	—	—	—	—	—	—	—	—
Sept. 5	14	0.81	0.66	0.79	0.64	0.70	0.35	—	—	—	—	—	—
Sept. 19	15	0.65	0.42	0.69	0.46	—	—	0.72	0.46	—	—	0.80	0.58
Oct. 3	16	0.72	0.52	0.71	0.48	0.59	0.26	—	—	—	—	—	—
Oct. 17	17	0.76	0.58	0.79	0.64	—	—	0.61	0.23	—	—	0.76	0.42
Nov. 1	18	0.72	0.50	0.75	0.55	0.42	0.06	—	—	—	—	—	—
Nov. 15	19	0.67	0.47	0.73	0.54	—	—	0.33	0.02	—	—	0.43	0.0
Nov. 29	20	0.54	0.43	0.82	0.70	0.07	0.0	—	—	—	—	—	—
Dec. 13	21	—	—	—	—	—	—	0.02	0.0	—	—	0.10	0.0

Table 19 — Probabilities for a Working Day

Suggested Readings

Agricultural Machinery Management. 1998. ASAE STANDARDS 1998. St. Joseph, Michigan.

Agricultural Machinery Management Data. 1998. ASAE STANDARDS 1998. St. Joseph, Michigan.

Borgman, D.E., and E. Hainline. 2008. Tractors (Fourth Edition). Deere & Company Service Publications. Moline, Illinois.

Bowers, W. 1994. Machinery Replacement Strategies. Deere & Company Service Publications. Moline, Illinois.

Breece, H.E. 1992. Planting (Third Edition). Deere & Company Service Publications. Moline, Illinois.

Buckingham, F., K.R. Carlson, and J.R. Conrads. 1991. Machinery Maintenance. Deere & Company Service Publications. Moline, Illinois.

Goering, C.E. 1992. Engine and Tractor Power (Third Edition). American Society of Agricultural Engineers, St. Joseph, Michigan.

Griffin, G.A. 2007. Combine Harvesting (Fifth Edition). Deere & Company Service Publications. Moline, Illinois.

Harrington, T.M., and C.A. Rotz. 1994. Draft Parameters for Major Implements. Paper No. 941533. American Society of Agricultural Engineers, St. Joseph, Michigan.

Hathaway, L., and F. Buckingham. 2007. Preventive Maintenance (Seventh Edition). Deere & Company Service Publications. Moline, Illinois.

Hunt, Donnell. 1995. Farm Power and Machinery Management (Ninth Edition). Iowa State University Press, Ames, Iowa.

Liljedahl, J.B., P.K. Turnquist, D.W. Smith, M. Hoki. 1989. Tractors and Thier Power Units (Fourth Edition). Van Nostrand Reinhold, New York.

Rider, A.R., and S.D. Barr. 2004. Hay and Forage Harvesting (Fifth Edition). Deere & Company Service Publications. Moline, Illinois.

Siemens, J.C., R.G. Hoeft, and A.W. Pauli. 1993. Soil Management (First Edition). Deere & Company Service Publications. Moline, Illinois.

Siemens, J.C., K. Hamburg, and T. Tyrrell. 1990. A Farm Machinery Selection and Management Program. Journal of Production Agriculture: 3:212:219.

Glossary of Terms

The following glossary of terms has been adapted from the ASAE Standard: ASAE S495—Uniform Terminology For Agricultural Machinery Management.

A

Accumulated Cost

Total cost for a period of time. In this text it is usually taken from the date of purchase. It can apply to all of the cost factors, or for individual costs. For example, it could apply to accumulated repair costs.

B

Brake Horsepower

The maximum power the engine can develop without alterations.

Breakdown

An unexpected change in duty status from operational to non-operational, due to mechanical failure.

C

Custom Cost

The amount paid for hiring equipment to perform a certain task.

Custom Rate

The rate of charge for a custom operation.

D

Depreciation

The reduction in value of a machine.

Drawbar Power

The pulling power of a tractor by way of tracks, wheels, or tires.

Drawbar Pull

The force exerted by the power unit at the drawbar to pull an implement.

E

Effective Field Capacity

Actual rate of performance of land or crop processed in a given time, based on total field time.

Effective Operating Width

The width over which the machine actually works. It may be more or less than the measured width of the machine.

F

Failure

The inability of a machine to perform its function under specified field and crop conditions.

Field Efficiency

Ratio of effective field capacity to theoretical field capacity.

Field Speed

Average rate of machine travel in the field during uninterrupted period of functional activity. For example, functional activity would be interrupted when the implement is raised out of the soil.

Field Time

The time a machine spends in the field measured from the start of functional activity to the time the functional activity in the field is completed.

Fixed Costs

Costs that do not depend on the amount of machine use, such as depreciation, interest on investment, taxes, insurance, and storage.

L

Lease

An contract that provides a method of obtaining the services of a piece of equipment in return for periodic payment. The period of a lease is normally considered as longer than one year.

Load Factor, Field

The ratio of engine power used in performing an operation to engine power available.

M

Maintenance

Cleaning, oiling, greasing, adjusting, etc., to keep a machine in operative condition and to help maintain its efficiency.

Major Overhaul

Extensive rebuilding that renews the efficiency of a machine and maintains its usefulness.

O

Obsolete

The condition of a machine either when it is out of production and parts to repair or update it are not available from normal suppliers or when it can be replaced by another machine or method that will produce a greater profit.

Operating Costs

Costs that depend directly on the amount of machine use.

P

Power-Take-Off Power

PTO power, in short, is the power measured at the PTO shaft.

R

Reliability

The ability of a machine to perform at its intended quality and capacity level during the time period that it is scheduled to operate.

Remaining Value

Sometimes referred to as "as-is" value. It is the fair market value of a machine, without a trade.

Rent

Similar to a lease, except the term of the contract is usually a year or less, for example, a day, hour, week, or month.

Repair

Restoring a machine to operative condition after breakdown, wear, accident, etc. Repairs do not add to the value or prolong the life of a machine. Major overhauls are not repairs.

S

Soil Resistance

The resistance of an implement as it is moved over or through the soil to accomplish the desired results.

Specific Fuel Consumption

The fuel consumed by an engine to deliver a given amount of energy. (Gallons per hour per PTO horsepower.)

T

Theoretical Field Capacity

Rate of performance obtained if a machine performs its function 100% of the time at a given operating speed using 100% of its theoretical width.

Theoretical Operating Width

The measured width of the working portion of a machine. For row-crop machines, it is the average row width times the number of rows.

Timeliness

Ability to perform an activity at such a time that quality and quantity of product are maximized.

Timeliness Cost

The reduction in production value due to inability to perform an operation during the time that quality and quantity of product are maximized.

Timeliness Factor

Sometimes called a coefficient, it is a factor representing proportional reduction in return per unit of time spent performing a given activity because of lack of timeliness.

Total Costs

The sum of fixed and operating costs.

U

Useful Life

The service life of a machine before it becomes unprofitable for its original purpose due to obsolescence or wear.

INDEX

A

Accidental Breakage or Damage 8-3
Acreage Chart A-29
Acres per Hour 2-2
Acres per Hour Capacity 2-3
Acres per Hour EFC 2-9
Adjustments 3-5
Alternatives, Comparing 13-3
Annual Costs 11-5

B

Brake Power 5-7
Breakdowns 3-5
Budgets
    Case Study 14-6

C

Calculating
    Lost-Time Costs 8-4
    Machine Life 10-7
Calculating Fuel and Lubricant
  Costs in Metric Units 7-9
Calculating Relative Costs 11-3
Capacity
    Unused 3-2
Capacity Measuring Methods 2-2
Case Studies
    Cash Flow Analysis 14-7
    Cost of New Machinery 14-4
    Financial Analysis 14-5
    Record Keeping 14-1
    Using a Budget 14-6
Cash Flow Analysis
    Case Study 14-7
Changing Operators 3-6
Checking Machine Performance 3-6
Combines
    Estimating Costs 9-8
Comparing Alternatives 13-3
Comparison of Four Alternative A-18
Converting Power Ratings 5-8
Cost for Individual MachinesCost
    Individual Machine 9-2
Cost Tables
    Machine Cost 9-2

Costs
    Estimating Chisel Plow 9-6
    Estimating Combine 9-8
    Estimating Cultivator 9-6
    Estimating Disk 9-6
    Estimating Harrow 9-6
    Estimating Mulch Tiller 9-6
    Estimating Tractor-Machine 9-6
Costs, System 9-9
Cropping Systems
    Comparing Fuel Consumption 7-4
Custom Work
    Annual Fixed Cost 12-3
    Costs 12-2
    Determining Average Operating Costs 12-4
    Determining Total Costs 12-4
    Establishing a Rate 12-5
Custom Work and Ownership 12-7
Custom Work to Reduce Costs 12-8

D

Declining Balance Depreciation 6-7
Depreciation 6-4
    Declining Balance 6-7
    Straight-Line 6-5
    Sum-of-the-Digits 6-6
Determining and Comparing Costs
    Custom Work 12-2
Determining Annual Fixed Cost 12-3
Determining Average Operating Costs 12-4
Determining Hectares A-30
Drawbar Performance A-22, A-23, A-24, A-25
Drawbar Power 5-8

E

Effective Field Capacity 2-9
    Area per Hour 2-9
    Estimating 4-2
    Weight per Hour 2-10
Energy and Fuel Requirements Table A-16
Engine
    Power Ratings 5-2
    Types 5-2
Equipment Life 8-5
    Estimating 8-5
Estimated Field Efficiencies 3-7

Estimating 8-6
    Average Fuel Consumption 7-5
    Chisel Plow Costs 9-6
    Combine Costs 9-8
    Cultivator Costs 9-6
    Disk Costs 9-6
    Effective Field Capacity 4-2
    Fuel and Lubricant Costs 7-7
    Fuel Consumption, Self-Propelled Machines 7-6
    Harrow Costs 9-6
    Lubricant Costs 7-7
    Machine Cost 9-4
    Mulch Tiller Costs 9-6
    Repair Costs 8-6
    Tractor-Machine Costs 9-6
Estimating Fixed Costs 6-11
Estimating Fuel Needs 7-2
Expansion, Allowing for 11-6

F

Field Capacity
    Effective (EFC) 2-9
    Measuring Methods 2-2
    Theoretical (TFC) 2-8
Field Conditions 3-3
Field Efficiencies
    Changing Operators 3-6
    Estimated 3-7
    Machine Performance 3-6
    Making Adjustments 3-5
    Reducing Breakdowns 3-5
    Rest Stops 3-5
    Servicing Machines 3-5
    Turning Time and Field Conditions 3-3
    Unclogging Machines 3-4
    Unloading Procedures 3-3
    Unmatched Machine Capacity 3-6
    Unused Capacity 3-2
Financial Analysis
    Case Study 14-5
Fixed and Repair Cost Tables A-2
    All Wheel-Type Tractors A-2
    Crawler Tractors A-3
Fixed Costs
    Depreciation 6-4
    Estimating 6-11
    How to Reduce 6-12
    Insurance 6-10
    Interest 6-10
    Low Annual Use 6-11
    Shelter 6-10
    Taxes 6-10
Fuel
    Types 7-4

Fuel Consumption
    Comparing Cropping Systems 7-4
    Estimating Average 7-5
    Estimating, Self-Propelled Machines 7-6
    Estimating, Tractors 7-5
Fuel Consumption Table A-15
Fuel Needs
    Estimating 7-2
    Horsepower-Hours of Energy 7-2
Fuel Saving Tips 7-8

G

Glossary of Terms A-34
Guidelines
    Machine Trade-in 10-3
    When to Repair 10-7
    When to Trade-in 10-8

H

Hectares per Hour 2-2
Horsepower-Hours of Energy 7-2

I

Implement
    Matching to Tractor 5-9
Insurance 6-10
Interest 6-10

L

Leasing 13-2
Lost-Time Costs
    Calculating 8-4
Low Annual Use Fixed Costs 6-11
Lubricant Costs
    Estimating 7-7

M

Machine
- Calculating Usefull Life 10-7
- Capacity 2-3
- Checking Performance 3-6
- Leasing 13-2
- Making Adjustments 3-5
- Obsolesence 10-6
- Renting 13-1
- Servicing 3-5
- Trade-in Guidelines 10-3
- Unclogging 3-4
- Unmatched Capacity 3-6
- Wear 10-6

Machine Cost
- Estimating 9-4
- Using Tables 9-2

Machine Reliability 10-6
Machine Size, Selecting 11-7
Machine Width 2-7
Machines and Power Units
- Matching 1-2

Managing Machinery 10-2
Matching Machines and Power Units 1-2
Material Capacity
- Measuring Methods 2-3

Measurement Conversion Chart A-28
Measuring Methods
- Capacity 2-2
- Field Capacity 2-2
- Material Capacity 2-3
- Throughput Capacity 2-3

N

Nebraska Tractor Test A-21

O

Obsolesence, Machine 10-6
Operating Speed, Selecting 2-4
Operator Neglect 8-4
Operators, Changing 3-6

P

Parameters for Estimating Repair Costs Table A-27
Power Rating, Engine 5-2

Power Ratings
- Brake 5-7
- Converting 5-8
- Drawbar 5-8

Power Requirements
- Tractor Sizes 5-7

Power Take-Off 5-7
Power Take-Off Horsepower
- Power Ratings 5-7

Probabilities for a Working Day A-32
Putting a Value on Timeliness 11-9

R

Record Keeping
- Case Study 14-1

Reducing Fixed Cost
- Ownership Time 6-13
- Proper Amount of Equipment 6-12
- Used Equipment 6-13

Relative Costs
- Calculating 11-3

Renting 13-1
Repair Costs 8-6
- Total Accumulated 8-8

Repairs, Types of 8-3
Rest Stops 3-5
Routine Overhauls 8-4
Routine Wear 8-3

S

Seeds or Plants per Acre A-30
Seeds or Plants per Hectare A-31
Selecting
- Machine Size 11-7
- Tractor Size 11-2

Servicing Machines 3-5
Shelter 6-10
Sizing for Critical Work 5-11
Soil Resistance 5-6
Soil Resistance Table A-15
Speed, Operating 2-4
Straight-Line Depreciation 6-5
Suggested Readings A-33
Sum-of-the-Digits Depreciation 6-6
System Costs 9-9

T

Taxes	6-10
Three Point Hitch Performance	A-26

Throughput Capacity
- Measuring Methods 2-3

Tillage Methods Table	A-15
Timeliness	11-7
Tire, Ballast and Weight	A-23
Total Accumulated Repairs (TAR)	8-8

Total Average
- Annual Costs 11-5

Tractor
- Selecting Size 11-2

Tractor Sizes
- Power Requirements 5-7

Tractors
- Estimating Fuel Consumption 7-5

Trading
- Average Cost per Unit 10-4
- Machine Obsolesence 10-6
- Machine Reliability 10-6
- Worn-Out Machinery 10-6

Trading Guidelines	10-3
Turning Time	3-3

U

Unclogging Machines	3-4
Unloading Procedures	3-3
Unmatched Machine Capacity	3-6
Unused Capacity	3-2

W

Wear, Machine	10-6
Weights and Measures	A-28
Wheel Slip, Determining	5-11
Worksheet for Comparing Alternatives	A-17